Social Dynamics in a Systems Perspective

New Economic Windows

More information about this series at http://www.springer.com/series/6901

Sergio Barile • Marco Pellicano •
Francesco Polese
Editors

Social Dynamics in a Systems Perspective

Springer

Editors
Sergio Barile
"Sapienza" University of Rome
Rome, Italy

Marco Pellicano
Salerno University
Salerno, Italy

Francesco Polese
Salerno University
Salerno, Italy

ISSN 2039-411X ISSN 2039-4128 (electronic)
New Economic Windows
ISBN 978-3-319-61966-8 ISBN 978-3-319-61967-5 (eBook)
DOI 10.1007/978-3-319-61967-5

Library of Congress Control Number: 2017949998

Printed on acid-free paper

This Springer imprint is published by Springer Nature
The registered company is Springer International Publishing AG
The registered company address is: Gewerbestrasse 11, 6330 Cham, Switzerland

Editorial: Social Dynamics in a Systems Perspective

The evolutionary path of science is traced along two distinct epistemologies: the naturalist one and the social one. The first is understood as the study of the existing through rigorous relations of causality, describing each phenomenon as the multiplicity of its components and of their relations; the second is oriented toward the comprehension of human phenomena through the contribution of sociology and social psychology.

However, over time, the consolidated analytical paradigm, positivist and based on the 'if so... then so' logic, has revealed its inadequacy, concluding that the representation of social phenomena through laws with similar "mathematical shape" is not only partial but also misleading.

Endowment and variety, interaction and organization offer, within the framework defined by systems thinking, a perspective rich in interdisciplinary implications: a social system, understood as finalized information variety toward a goal, is influenced by the principle of consonance, moving toward increasing levels of resonance, while the same system, analyzed through the metrics of its performance, is understood as structure, allowing precise descriptions of the same variety and highlighting the relationships among its components. More related to the first representation are the studies of social, psychological, and humanistic sciences: the emphasis is on different actors' behavior and on their interaction in a focalized context.

The hard sciences, instead, describe the same dynamics through the phase transition process, the transformation, and the maintenance of explanatory characteristics. This expressed continuity is well summarized by an always-in-evolution, robust body of theories that has allowed the definition of methodologies, techniques, and tools of proved scientific efficacy. In this perspective, the systemic representation emphasizes the emergence of a collective scientific mind able to bind together and to combine the different specificities of knowledge under a shared general scheme of construction.

In this scenario, it appears desirable to define a common approach to the study of apparently distinct and distant—with respect to the applicable metrics—

phenomena: previous economies of thought, built on similarities and on forced conceptual postponements, are no longer suitable to the understanding of living systems.

The question, therefore, can only be bound to the need to fill a conceptual vacuum: how to qualify the divergence in terms of representation and description between living systems (viable) and nonliving systems? What distinguishes a social system from a physical system? What distinguishes a physical system from the individuals as systems?

A closer look, in fact, reveals that the principle of equifinality is not sufficient to mark the groove. Yet, in affirming its failure, no law and no general pattern apt to describe the developmental trajectories of social systems appear likely to be uniquely defined: it always remains an ignored condition, an irreducible quantity, or an unavoidable assumption of simplification or indeterminacy. Life and, in general, social phenomena simply appear beyond the resolutive possibility of the analytical method. Despite this consideration, living and nonliving systems share a basic ontology that covers them simultaneously (cybernetics is a clear example): the centrality of information and the way in which it is organized and processed. This volume aims to provide a clearer evidence of systems thinking multiplicity and of its wide applicative heterogeneity. The contributions come from different conceptual matrices, ranging from the study of social dynamics, through the concepts of sustainability and value co-creation, to the design of complexity and from the analysis of crucial managerial dynamics (innovation is an example) to the definition of decision-making processes as indispensable lever of each managerial logic, until the possibility to imagine the occurrence of extremely rare and high-impact events (black swans). Given these premises, it does not appear appropriate to promote research aimed to quantify the amount of information, but it becomes fundamental to understand the role played by each individual in selecting his relevant context and the individuality of knowledge itself, according to the principles of creativity and variety. Although it is not possible to extend the specificity of each field of research in each area of knowledge, it is certainly possible (and even desirable) to understand that any process of acquisition and generation of knowledge is linked to general laws that, through a common and coordinated action, have inspired the creation of each science. Certainly, the detection of baseline cultural areas, such as complexity and constructivism, system dynamics and innovation, sustainability and value co-creation, does not qualify as conceptual watershed but allows us to justify similar paradigmatic affinities and thus the adoption of unique, specific code of language: although centered on diverse themes, the contributions briefly described below are all related to the contribution that the systemic epistemology, mobile and constructivist, provides to the reading of social phenomena.

In this direction, Tronvol, Barile, and Caputo stress the importance of a paradigm shift in marketing studies, due to the profound changes that have characterized the postmodern era. Social phenomena and social structures influence consumer choices: therefore, a renewed interpretation of the traditional cause–effect logic, wearing the lens provided by the systems thinking, is needed.

Moreover, systems thinking approach deeply connotes the intervention of Polese, which depends on the concept of value co-creation in the theoretical framework provided by Viable Systems Approach (ASV). In this light, Polese emphasizes the "behavioral dimension" of the integration among actors and resources not only in terms of structural compatibility (consonance) but also in terms of will of interaction (resonance).

Accordingly, Barile and Saviano reconstruct the epistemological premises at the base of Viable System Approach, placing a strong emphasis on the growing inadequacy of traditional "management toolkits" for dealing with complexity and on the centrality of the relationship between vitality and sustainability.

The work of Faggini and Parziale explores the contribution that systems thinking provides to the understanding of the pair innovation–growth, highlighting the role of institutions and the relevance of coordination policies addressed toward the understanding of complexity.

Afterward, Walletzky, Buhnova, and Carrubbo investigate the relationship between Smart City and Smart Citizen, overcoming a perspective solely centered on the role of service, paying attention to the needed engagement of individuals to define an effective process of value co-creation.

In this direction, Löbler reads the concept of service as an entropy reduction factor, directed to create the conditions for a sustainable coexistence of humans and natural agents.

Espejo proposes an interpretative model to understand rare but recurrent events, known as black swans, and suggests the existence of a general scheme that binds and covers the set of social phenomena as belonging to early common roots.

The contribution of Botti, Grimaldi, Tommasetti, and Vesci is part of a discussion of crucial importance in Service-Dominant Logic paradigm and, in general, in managerial sciences: the issue of measurability of co-created value.

A different approach is used by Sigala, who analyzes the co-creation process in its relationship with the sharing economy, studying the Airbnb phenomenon, while Pellicano, Ciasullo, Troisi, and Casali emphasize, in a systems perspective, the relational and interactional nature of each process of value co-creation.

Further on, Mele and Russo-Spena bind together the effectiveness of a value proposition to the values of the different stakeholders involved in the process, underlining how actors' values, aims, and practices shape a service ecosystem, impacting on its viability.

Calabrese, Iandolo, Caputo, and Sarno offer a comprehensive review of existing literature on systems thinking, summarizing its limitations and issues and highlighting their impact on decision-making processes.

In conclusion, Bruni, Carrubbo, Cavacece, and Sarno examine the systems thinking contribution to the understanding of the dynamics that define and tie together marketing and management.

Without any claim of exhaustiveness, it is our fervent hope that the attentive reader to systems thinking evolution could find in these contributes a valuable tool for his cultural commitment and future researches.

Contents

Chapter 1
A Systems Approach to Understanding the Philosophical Foundation of Marketing Studies

Bård Tronvoll, Sergio Barile, and Francesco Caputo

Abstract Marketing represents a multidisciplinary research domain in which several instruments, models and approaches have been provided to both researchers and practitioners in order to better face the emerging social and economic challenges. Despite the relevance of all contributions provided by marketing studies, the academic marketing domain seems to be affected by a strong reductionist view in which the focus is on the single actors and transactions as key links between companies and environments. With the aim to overcome the limitations of reductionist view in academic marketing studies, this chapter adopts the interpretative lens provided by systems thinking. The ontology and epistemology of academic marketing studies are analysed in light of the systems perspective, and a transcendent philosophical perspective is proposed to support the building of more effective multi- and trans-disciplinary approaches and models within the marketing domain.

Keywords Systems Approach • Marketing • Philosophical Foundation

1.1 Introduction: Hazards and Opportunities of Market Exchanges in the Postmodern Era

The technological and economic developments of the last century have enhanced the change and evolution of social and economic dynamics, influencing the social balance and pushing the emergence of new social structures built on new roles (Toffler and Alvin 1981; Perez 1983; Inglehart 1997). Over the past decades, markets and social configurations have shown increased incapability in

B. Tronvoll
Inland Norway University of Applied Science, Elverum, Norway

S. Barile
Sapienza, University of Rome, Rome, Italy

F. Caputo (✉)
Masaryk University, Brno, Czech Republic
e-mail: fcaputo@mail.muni.cz

S. Barile et al. (eds.), *Social Dynamics in a Systems Perspective*, New Economic Windows, DOI 10.1007/978-3-319-61967-5_1

understanding the increasing social and economic variety. This leads to the need for more efficient, effective and sustainable managerial models has become a new challenge to face (Holling 2001; Miller and Page 2009).

The changes in market and social dynamics have been influenced by key events such as the altered social and psychological boundaries in market relationships (Dunfee et al. 1999). Furthermore, the opportunities offered by the new models are built on peer-to-peer flows, and the increasing knowledge and awareness of social and economic actors has facilitated change (Palmer and Ponsonby 2002; Plouffe 2008; Caputo et al. 2016a, b). All of these events affect the relationships and variables that are able to influence companies' strategies and survival (Akaka et al. 2013). Traditional approaches built on social and hierarchical market configurations have been progressively altered, and new market dynamics have emerged (Tadajewski and Jones 2009; Goldfarb et al. 2012). More specifically, the traditional transactional and deterministic approaches to production have changed because they are built on standardisation, and the key parameters are not aligned with the needs of emerging markets (Roberts 2000). Thus, previous models were based on a top-down approach, and companies' controlled flows of information have now been replaced by new models in which every actor (e.g., customers, employees, and companies) can provide information that influences market perceptions and behaviours (Tadajewski 2004; Cova and Dalli 2009; Barile et al. 2015a, b, c; Evangelista et al. 2016; Caputo and Walletzký 2017).

The proposed changes in social and economic dynamics can be considered partial examples of the whole evolutionary dynamic summarised by the emergence of a new time called the 'postmodern era' in which 'the immutability of authority is challenged—not only by "science" but, more immediately, by the mundane expansion of market relations and the distinctive capabilities of labour processes' (Wilmott 2001, p. 645). As a consequence of multiple vibrant dynamics that involve every kind of actor in a complex situation, it is not possible to clearly define boundaries and rules and in which dynamics, balances and structure continually evolve with time. In accordance with this, the essence of emerging markets and social configurations imposed by the postmodern era is summarised by Giddens (1991), stating, 'the reflexivity of modernity actually undermines the certainty of knowledge. The integral relation between modernity and radical doubts is an issue which, once exposed to view, is not only disturbing to philosophers but is *existentially troubling* to ordinary individuals' (p. 21). In other words, the postmodern era has shattered old balances, outlining the need for new dynamics and social structures. As a result, traditional approaches based on reductionist and cause-effect logic must be challenged by wider and holistic approaches that include long-forgotten variables and dimensions (Fuat Firat and Shultz 1997).

In this emerging scenario, academic marketing studies must face challenges imposed by the increasing uselessness of traditional marketing approaches, because they cannot simply influence consumers' preferences and behaviours. Academic marketing need to identify new pathways and strategies through which companies can build direct links with their constituents to better understand and satisfy their needs and expectations (Tronvoll 2012). A paradigmatic change is required in academic marketing with the aim to deeply analyse the relevant cognitive and psychological dimensions on which the market is based, in order to support the emergence of a holistic marketing approaches and research pathways (Maclaran 2009).

The remainder of this chapter will discuss how to achieve this aim, and is structured as follows: Sect. 1.2 proposes a brief analysis of multi- and trans-disciplinary methods on which academic marketing studies are based; in Sect. 1.3, some preliminary reflections are presented with references to the definition of systems-thinking-based ontology and epistemology in the philosophical foundation of academic marketing studies; Sect. 1.4 discusses possible implications related to a wider understanding of systems thinking implications in marketing studies; Sect. 1.5 highlights the need to adopt the proposed extension in perspective-defining recursive and dynamic pathways; and, finally, in Sect. 1.6, some final remarks and possible future lines of research are presented.

1.2 The Need for a Wider Perspective in Marketing Studies

Bartels (1976) claimed 'the essence of marketing was the combination of factors. Blindness to and ignorance of that combination of factors is the reason for the absence of terms equivalent to marketing in other languages. Marketing must be regarded not merely as a business practice, but as a social institution' (p. 13). With this reflection in mind, the author underlined the multi-dimensions on which marketing is based and indirectly highlighted the need to build approaches and models that are able to catch this peculiarity of marketing studies and define pathways in which the multiple dimensions of marketing are clearly linked in a common conceptual framework (Bartels 1976).

Recent calls for multi- and trans-disciplinary approaches to academic marketing studies remain unanswered (Moulaert 2013). Different authors (e.g., Hunt 2002; Golinelli et al. 2012) have underlined the need to overcome the reductionist business view of marketing and better define sociological and psychological foundations in academic marketing studies. More specifically, Doyle and Stern (2006) underlined that marketing is based on the understanding of peoples' needs, perspectives and perceptions; therefore, it cannot simply bend to individual companies' aims and strategies. Following the same pathway, Hunt (1994) outlined the need for rethinking logic and purposes in academic marketing studies in order to extend the perspective on which they are based. In the same direction, Lemmink (2005) underlined that 'based on the intrinsic qualities of theories and models from our own disciplines like marketing and operations management, we should be able to put more effort into collaboration' (p. 8).

All of these contributions represent evidence for a lack of knowledge in the research community toward gaining a more complete understanding of the marketing domain (Kohli and Jaworski 1990). This condition can be explained as the consequence of a growing specialisation in marketing studies, which is connected to the increasing incapability to link multiple hyper-specialised marketing contributions within a common interpretative and conceptual framework (Alderson 2006).

According to Aaker (2008), it is possible to underline that the current approach to academic marketing studies is based on the definition of multiple 'silos' that are strictly related to specific domains, as well as research areas that are apparently linked by their shared view of marketing based on a few general principles. This approach offers the advantage to build detailed knowledge about specific domains, but it is affected by its incapability to effectively support the explanation and management of social and economic dynamics (Achrol 1991).

To bridge this gap, one must overcome the limitations imposed by a reductionist view toward marketing studies and build extended knowledge about the foundation on which marketing is based. For this purpose, the following sections offer some reflections on current opportunities related to the building of a dynamic and systems view of marketing.

1.2.1 From a Mechanical to a Dynamic View Toward Academic Marketing Studies

Over the last few years, an increasing variety of studies has been requested in order to produce more efficient approaches that are able to support decision makers in facing the emerging challenges (Srivastava et al. 1999; Edvardsson et al. 2011b). In answer to this need, many researchers have tried to apply the principles and guidelines of the 'strong' sciences, such as physics, mathematics and biology, to the social domains (Bohman 1993). The intention of this approach is to identify the possible principles on which the definition of a strong discipline able to investigate social dynamics, such as physical dynamics, is based (Verschuren 2001). In this direction, a great number of contributions have been offered with the intention to deeply investigate every possible dimension of social dynamics in order to build a strong knowledge base for this piece of the hypothetical puzzle (Rabinow 1987).

This increasing attention on the specific dimensions of social phenomena have influenced the emergence of multiple research streams directed toward building more specialised knowledge also in marketing studies (Redman et al. 2004). As a result, each research stream has developed its own languages, instruments and approaches (Kline 1995). The final step of this 'regressive pathway' was the building of specialised knowledge divided by impenetrable boundaries, in which a small group of actors attempt to define possible interpretative models through the identification of cause-effect relationships; however, this neglects the fact that all analysed elements are influenced by and, at the same time, influence other dimensions that are not considered and represented in the proposed interpretative models (Kauffman 2007).

All of these reflections can be summarised through an outline of a mechanist approach toward the social sciences, and specifically toward marketing studies, based on the definition of possible advancements in knowledge that are not very useful in understanding and managing social dynamics, because they are based on a partial representation of reality (Byrne 1998).

To bridge this gap, the potential pathway must overcome boundaries among the different research streams that are involved in exposing the marketing domain to a multi- and trans-disciplinary contamination, and must be oriented so as to observe the evolution of social phenomena over time and build possible explanations about their dynamics, not only useless descriptions of their structural compositions (Möller and Halinen 2000). Under this task, the real challenge in academic marketing studies is the identification of a shared ontology and epistemology able to reunite different possible approaches that are interested in the understanding of social dynamics under a common interpretative umbrella (Easton 2002).

1.2.2 *The Need for a Systems View in Academic Marketing Studies: Contributions from the Viable Systems Approach*

The need for rethinking traditional transactional approaches in academic marketing studies is an imperative, because a better understanding of market phenomena is vital to ensure the survival of social and economic systems (Fodness 2005). To such a degree, different research streams have underlined the need to improve instruments and interpretative models so that they are able to connect the different variables involved in marketing processes and dynamics in a holistic, interpretative approach, investigating the ways in which different elements interact to achieve a specific aim (Bettis and Prahalad 1995).

Current approaches based on the investigation and explanation of phenomena are showing increasing uselessness, because they require more time to identify and link key variables, test possible relations and define possible managerial instruments (Brown and Eisenhardt 1997). In the time that traditional approaches need to understand a particular phenomenon, more evolutions in social and economic dynamics have occurred and the investigated phenomena become outdated before a clear explanation can be formulated (Stacey 1995).

To overcome this limitation, the potential pathway must shift the attention from the evidence of social and economic dynamics to the elements and variables that influence them (Sarasvathy 2001). More specifically, it must identify and investigate the elements that influence the emergence and evolutions of social and economic dynamics over a period of time, overcoming the strict focus on 'tangible evidence' (Barile et al. 2012).

A possible contribution is offered by systems thinking as a multi- and trans-disciplinary interpretative approach that focuses on the ways in which a set of individuals, companies, knowledge and processes interact in a shared environment through a complex network of interrelationships affected by divergent aims, perspectives and perceptions (Beer 1979; Capra 1985; Espejo 1994; Wolf 1999; Golinelli 2010; Barile and Saviano 2011). According to Werhane (2008), the systems view outlines that 'almost everything we can experience or think about is in a network of interrelationships such that each element of a particular set of interrelationships affects some other components of that set and the system itself, and almost no phenomenon can be studied in isolation from other relationships with at least some other phenomenon' (p. 467).

Building on this definition, it is possible to highlight that the interpretative lens offered by systems thinking supports the understanding of interactions among different approaches, variables, perspectives and bodies of knowledge within the context of social and economic dynamics (Barile and Saviano 2010; Mele et al. 2010; Saviano et al. 2016). The systems view offers the opportunities for overcoming boundaries and distances among different research streams and pathways involved in the academic marketing domain, underlining the need to shift attention away from the specific dimensions investigated and toward the principles and logic that influence them (Webster and Lusch 2013). The adoption of a systems view in marketing studies will help identify and investigate a possible shared philosophical foundation that is able to combine the different 'marketing silos' into a holistic and shared conceptual framework (Vargo and Lusch 2011).

Adopting the systems view in marketing studies means overcoming the functional view of marketing as a complex of company activities and defining a more efficient solution for supporting the relationships between companies and context (Lusch and Vargo 2014; Saviano et al. 2014). Systems logic underlines the social role of marketing as a complex of models, approaches and instruments that support a better understanding of the ways in which social, economic and psychological dimensions interact in building everyday dynamics and balances (Smith et al. 2010).

In accordance with the proposed reflections, a useful interpretative contribution is offered by the Viable Systems Approach (VSA) as a meta-level model that is able to link different disciplines and research streams through a common pathway, which will support the understanding of every kind of organised phenomenon such as companies, communities, families and others (Barile 2009; Golinelli 2010; Barile et al. 2015a, b, c). The VSA represents a bridge between the reductionist view and a more holistic approach, supporting the shift from focusing on the elements that compose a system (i.e., the structure) to the ways in which the elements interact in order to achieve a common aim (Golinelli et al. 2012).

With the aim of supporting the understanding and management of complex social and economic phenomena, the VSA proposes to observe any organisation as a 'viable system' that can survive by interacting with other systems, which are subjectively classified with reference to their abilities to impact on the organisation's viability (Barile et al. 2012).

Building on this key concept, the VSA identifies 10 fundamental concepts that can support the emergence of a systems view in marketing studies (see Table 1.1).

Table 1.1 The 10 fundamental concepts (FCs) of VSA

FC1	Individuals, organizations, and social institutions are systems that consist of elements directed towards a specific goal
FC2	Every system (of level L) identifies several supra-systems, positioned at a higher level (L+1), and several sub-systems, located at a lower level (L−1)
FC3	The interpretation of complex phenomena requires interdisciplinary approaches and should synthesize both a reductionist view (analysing elements and their relations) and an holistic view (capable of observing the whole)
FC4	Systems are open to connection with other systems for the exchange of resources. A system boundary is a changing concept within which all the activities and resources needed for the system's evolutionary dynamic are included
FC5	Viable systems are autopoietic and self-organizing; that is, they are capable of self-generating internal conditions, which through self-regulation, support the reach of equilibrated conditions, thus synthesizing internal possibilities and external constraints
FC6	Every organization is constituted by components that have specific roles, activities, and objectives, which are undertaken within constraints, norms, and rules. From structure emerges a system through the transformation of relations into dynamic interactions with sub-systems and supra-systems
FC7	Systems are consonant when there is a potential compatibility among the system's components. Systems are resonant when there is effective harmonic interaction among components
FC8	A system's viability is determined by its capability, over time, to develop harmonic behaviour in sub-systems and supra-systems through consonant and resonant relationships
FC9	Business dynamic and viability require continuous structural and systemic changes focused to the alignment of internal structural potentialities with external systemic demands
FC10	Viable systems continuously align internal complexity with external complexity in order to better manage changes affecting its viable behaviour. Decision-makers within these cognitive processes are influenced by strong beliefs, his/her interpretational schemes, and information

Source: Barile et al. (2012, p. 67)

1.3 Toward Systems-Thinking-Based Ontology and Epistemology in the Philosophical Foundation of Academic Marketing Studies

Building on the ideas proposed in previous sections, it is possible to state that the transactional view of marketing, as a complex of models, strategies and instruments with the specific aim of influencing social and economic dynamics in which actors are considered simple elements, cannot be considered useful to ensure the survival of organizations in the emerging social and economic dynamics (Vargo and Lusch 2004; Badinelli et al. 2012).

The vibrant society in which we live requires an in-depth investigation of the basic logic behind marketing in order to better identify and understand emerging social and market challenges (Pride and Ferrell 2016). To such an end, the following question requires an investigation: In what ways can the ontology and epistemology of academic marketing be rethought in order to be better aligned with social and economic dynamics? More specifically, considering that ontology draws attention to the identification and study of categories that either exist or may exist within a specific domain. Furthermore, epistemology refers to the study of the ways in which knowledge emerges and is used within a specific domain, the challenges of academic marketing studies requires an investigation of the ways that actors perceive social and economic dynamics (Gettier 1963; Griswold 2001; Peters et al. 2013).

Over time, marketing studies have been influenced by the evolution of social and economic dynamics and have attempted progressive definitions of possible pathways that may help explain environmental evolutions (Webster 2005). After Kumar (2015) proposed an initial definition for transactional-based marketing studies as supporting the selling of companies' products and service, increased attention has been paid on the ways in which relations and interactions among the actors involved in social and economic dynamics can influence the survival of every kind of organisation (Hills et al. 2008; Caputo 2016). Many researchers began underlining a need to investigate the social and psychological dimensions that influence actors' perceptions, decisions and behaviours (Furrer et al. 2000). Some authors outlined that it is not possible to define an invariant and shared approach in marketing studies because social and economic phenomena are perceived in different ways by different actors (Tsai 2005).

In accordance with this, the academic marketing perspective must be extended, building upon contributions offered by psychological and philosophical studies that are interested in the analysis of individual and collective approaches to the interpretation of the environment (Lemmink 2005). A relevant advancement in knowledge could be offered by studies on constructivism, which, in contrast to determinism, affirm that individual and collective actors perceive the environment in which they live as a consequence of subjective and emotional meaning-making processes (Glasersfeld 1996). More specifically, theories based on the constructivist approach identify five themes of interest: (1) active agency, (2) order, (3) self, (4) social-symbolic relatedness and (5) lifespan development, as defined in Table 1.2 (Mahoney 2004, p. 363).

In analysing these themes, it is reasonable to state that social and economic actors cannot be considered individual entities and that social and economic dynamics emerge from the interactions between and reciprocal influences among social and economic actors. It can also be said that the ways in which actors define their pathways and strategies depend upon their capabilities to perceive the environment in which they live.

All of these reflections show that there is an incapability in investigating and understanding social and economic dynamics using the transactional approach, on which current academic marketing studies are based (Bryman and Bell 2015). In order to overcome this gap, systems thinking appears to offer a fruitful path. The opportunities to investigate each phenomenon as the result of a complex network of interactions between actors, resources and institutions, influenced by specific aims and perspectives, can offer new stimuli to marketing studies (Rubenstein-Montano

Table 1.2 Themes of interest in constructivist approach

Themes on interest	Core concepts
Active agency	"Human experiencing involves continuous active agency. This distinguishes constructivism from forms of determinism that cast humans as passive pawns in the play of larger forces"
Order	"Much human activity is devoted to ordering processes—the organizational patterning of experience by means of tacit, emotional meaning-making processes"
Self	"The organization of personal activity is fundamentally self-referent or recursive. This makes the body a fulcrum of experiencing, and it honors a deep phenomenological sense of selfhood or personal identity. But the self is not an isolated island of Cartesian mentation. Persons exist and grow in living webs of relationships"
Social-symbolic relatedness	"Individuals cannot be understood apart from their organic embeddedness in social and symbolic systems"
Lifespan development	All individual and collective actions "reflects an ongoing developmental flow in which dynamic dialectical tensions are essential. Order and disorder co-exist in lifelong quests for a dynamic balance that is never quite achieved. The existential tone here is unmistakable"

Source: Mahoney (2004, p. 363)

et al. 2001). Systems thinking proposes a shift from the hierarchical approach, in which identifiable causes influence and facilitate market dynamics, to a perspective in which market dynamics are the consequences of actors' will and attempts to combine their perspectives and aims into possible shared paths that are useful for achieving individual aims (Folke et al. 2005).

In some ways, these opportunities have been cached by recent evolutions in academic marketing studies as part of the value co-creation process and ecosystem dynamics (Vargo et al. 2008; Edvardsson et al. 2011a, b; Akaka et al. 2012; Dominici et al. 2017). Despite this, the majority of contributions offered that are related to these topics appear to be strictly oriented toward defining a possible explanation for market results, exclusively focusing on the interactions among actors and paying less attention to the motivations and logic behind these interactions (Caputo et al. 2016c). Recognising that there is a need to validate recent contributions toward academic marketing studies, there must be a better investigation of the sociological and philosophical grounds on which the increasing relevance of market relationships is based. Moreover, identifying the possible pathways that link different disciplines in a shared multi- and trans-disciplinary framework able to offer a holistic view about the variables involved in marketing studies.

This widening of perspectives can be supported by the conceptual and theoretical contributions offered by the VSA, as defined in the previous section. Adopting the interpretative lens offered by the VSA, it is possible to state that the emergence of systems-thinking-based ontology and epistemology in marketing studies requires that traditional approaches based on specialisation and the examination of individual elements and dimensions must be overcome. To build a more efficient view in academic marketing studies, it is necessary to clarify that it is not possible to investigate social and economic phenomenon in an objective way, because they have different features when observed by different entities (Barile et al. 2016).

1.4 A Transcendent Philosophical Perspective

As underlined by Hochschild (1979), ‘part of what we refer to as the psychological effects of “rapid social change”, or “unrest”, is a change in the relation of feeling rule to feeling and a lack of clarity about what the rule actually is, owing to conflicts and contradictions between contending sets of rules” (pp. 567–568). Based on this statement, the general approach to social studies and the specific view in academic marketing studies should identify new possible interpretative pathways (Tushman and Romanelli 2008). Approaches based on transactional logic and the achievement of individual aims alone cannot explain the complex dynamics that affect social and economic phenomena (Granovetter 2005; Gummesson and Polese 2009). While they are useful in describing and investigating tangible evidence related to market relationships and interactions, these elements can only be considered the tip of the iceberg, and the foundation is not considered by many approaches to marketing studies (Ferrell and Gresham 1985). To build a holistic interpretative framework, a ‘transcendent approach’ must be defined and investigated, able to link market evidence with the hidden and imperceptible variables related to sociological and philological dimensions representing the base of ‘market iceberg’. While market evidence can acquire different structures that evolve with time, such as the visible part of an iceberg, the fundaments are stronger and many evident changes in market dynamics could be explained via a clear knowledge of the invisible dynamics (Mavondo and Farrell 2003; Del Giudice et al. 2016).

The need for a transcendent approach in academic marketing studies does not require a simple replacement of old approaches with new ones; it requires an understanding of social dynamics through the identification of the rules and laws on which they are based. In this perspective, a useful advancement is offered by systems thinking in the form of a possible multi- and trans-disciplinary interpretative lens that is able to link different research streams and approaches within the academic marketing domain. Thanks to the systems view, it is possible to build a common umbrella in academic marketing studies through the identification of key principles that are related not only to static representations of social and economic phenomena, but also their evolutions over time (Barile et al. 2012). These evolutions include consequences of the combination and reciprocal influences of sociological, philosophical and psychological dimensions that affect the ways in which actors (both in individual and collective forms) perceive the world in which they live, define their aims, choose strategies to adopt and pathways to follow, and define the levels of relevance among resources they own, relationships in which they are involved and which of their strong beliefs to respect (Polese 2009; Di Nauta et al. 2015).

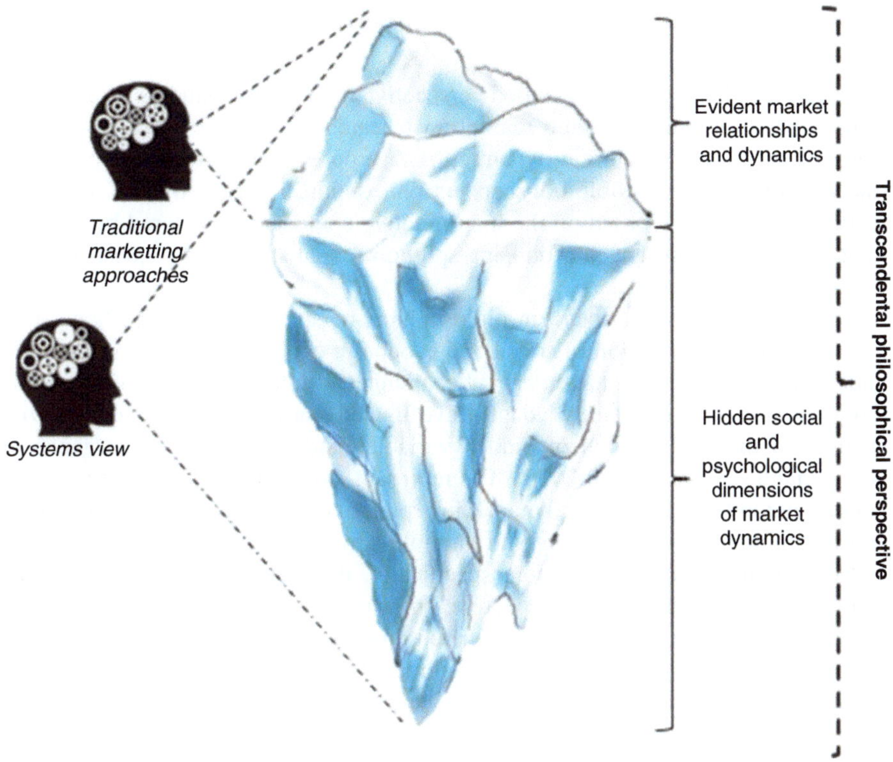

Fig. 1.1 The market iceberg

Following this line of thought, a real advancement in academic marketing studies requires the definition of a possible skeleton for a transcendent philosophical perspective able to combine evident and hidden variables within market dynamics, thanks to the interpretative contribution offered by the systems view. As shown in Fig. 1.1, to enrich traditional marketing approaches using the interpretative contributions offered by the systems view, with reference to the hidden variables on which the evident market relationships are based, means defining a transcendent philosophical perspective able to support marketing studies with a holistic view of social and economic phenomena. Thanks to this 'contamination', it is possible to easily adapt marketing instruments to social and economic evolutions and, at the same time, build more efficient instruments that are able to identify the relevant elements for perceiving possible future evolutions of the market.

1.5 The Root of a Recursive Perspective

The need for identifying an extended perspective in marketing studies is one of the most debated among academic marketing researchers who are interested in the study of social and economic dynamics. The proposed reflections and pathways traced in defining a possible transcendent philosophical perspective on academic marketing studies are based on the interpretative contributions offered by systems thinking, which represents a promising solution for satisfying the requests of both researchers and practitioners.

The adoption of a transcendent philosophical perspective opens the 'gate' to a more holistic interpretation of social and economic evolutions but, to be useful in understanding and facing the emerging challenges, it must be based on a dynamic representation that constantly adapts investigated variables in accordance with changes in the combination of hidden variables and the possible emergence of new variables (Letaifa et al. 2016). The representation in Fig. 1.2 is proposed as a basic skeleton for improving a dynamic interpretative model that is able to combine and recombine resources, actors and variables involved in changes within social and economic environments. More specifically, the identified variables must be analysed using a recursive perspective, in which marketing and philosophical foundations are related through their shared process of evolution.

Thanks to the adoption of the proposed recursive approach, it can be stated that the tangible evidence of the market (e.g., exchange, needs, social and economic actors, etc.) is the consequence of interactions and co-evolutions among the hidden philosophical pillars of marketing (e.g., the constructivist approach, individual and collective beliefs, and individual perceptions and interpretations). Accordingly, the definition of marketing approaches that are able to face emerging social and economic challenges can be applied to both marketing practitioners and

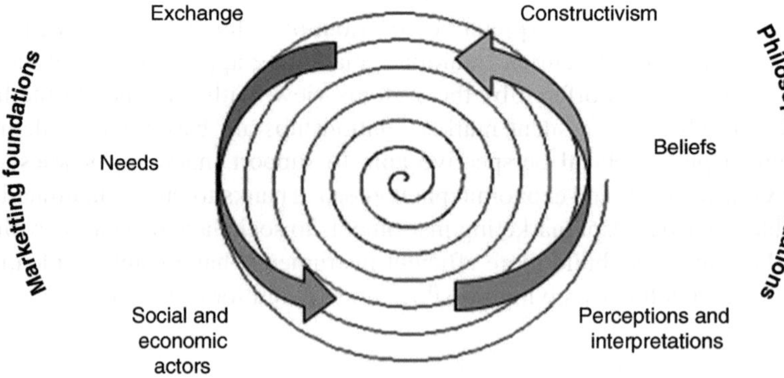

Fig. 1.2 A recursive perspective between marketing and philosophical foundations

researchers, extending their perspectives so that they may perceive the hidden dynamics of the market and formalise efficient instruments to manage its tangible manifestations.

Building on this recursive perspective, it is possible to define dynamic market models able to support the elaboration of interpretative approaches that are aligned with the changes in and evolution of social and economic dynamics. More specifically, starting from the philosophical foundations of market relationships and dynamics, it is possible to define an efficient, effective and suitable interpretative lens that is able to read each dimension of the marketing foundation holistically.

1.6 Conclusions

Recent changes in social and economic balances have resulted in the call for rethinking approaches and underlying concepts that represent, analyse and explain market dynamics and relationships. The traditional reductionist approach, based on cause-effect logic, shows an increasing uselessness because of its narrow focus on individual dimensions and its building upon a simple analysation of evident and tangible elements within market dynamics.

This chapter introduced current opportunities in extending academic marketing studies via the interpretative contributions offered by systems view. Academic marketing includes the hidden sociological and psychological dimensions on which market dynamics are based, as well as through the definition of holistic representations that are useful in improving efficiency, effectiveness, and sustainability in marketing approaches (Morgan et al. 2002).

This chapter proposed the adoption of a transcendent philosophical perspective in academic marketing studies, which will advance existing knowledge because it offers a possible solution for combining different disciplines and research pathways and defines a possible answer for multi- and trans-disciplinary models within the field of marketing (Moulaert 2013).

The adoption of a more holistic view in academic marketing studies via the definition of dynamic and recursive pathways, as shown in Fig. 1.2, represents the base for more relevant implications from both a theoretical and practical viewpoint. From a theoretical perspective, there are emerging needs to:

1. Better investigate social and psychological dimensions on which market dynamics and relationships are based (Tronvoll 2011);
2. Analyse the network of relationships among actors, resources and institutions in the light of the proposed transcendent philosophical perspective;
3. Identify possible conceptual and empirical instruments for measuring the direct and indirect influences of social and psychological dimensions on market relationships and dynamics (Patterson and Spreng 1997).

From a practical viewpoint, there are possible needs to:

1. Acquire and analyse data on market dynamics and trends in order to identify possible elements useful for understanding the social and psychological dimensions (Fabrigar et al. 1999);
2. Identify possible pathways and best practices, based on the analysis of social and psychological dimensions, for understanding and managing market relationships (Barile et al. 2015a);
3. Improve multi- and trans-disciplinary contaminations in marketing activities and strategies by better including social, cognitive, and psychological dimensions (Barile et al. 2015b).

While this chapter does not provide an exhaustive answer to the underlined need for a philosophical foundation in marketing studies, it offers a possible starting point on which future lines of research can be built to better support the contamination among existing approaches and models within the marketing domain and the opportunities related to a systems view of the market. Further research will focus on the possible formalisation of the role that the philosophical foundation of academic marketing studies can provide, including the definition of marketing models able to more suitably support solutions for the coping with emerging social and economic dynamics.

References

Aaker, D. A. (2008). Marketing in a silo world: The new CMO challenge. *California Management Review, 51*(1), 144–156.

Achrol, R. S. (1991). Evolution of the marketing organization: new forms for turbulent environments. *The Journal of Marketing, 55*(4), 77–93.

Akaka, M. A., Vargo, S. L., & Lusch, R. F. (2012). An exploration of networks in value cocreation: A service-ecosystems view. *Review of Marketing Research, 9*(Special Issue), 13–50.

Akaka, M. A., Vargo, S. L., & Lusch, R. F. (2013). The complexity of context: A service ecosystems approach for international marketing. *Journal of Marketing Research, 21*(4), 1–20.

Alderson, W. (2006). The analytical framework for marketing. In B. Wooliscroft, R. D. Tamilia, & S. J. Shapiro (Eds.), *A twenty-first century guide to Aldersonian marketing thought* (pp. 61–73). New York: Springer US.

Badinelli, R., Barile, S., Ng, I., Polese, F., Saviano, M., & Di Nauta, P. (2012). Viable service systems and decision making in service management. *Journal of Service Management, 23*(4), 498–526.

Barile, S. (2009). *Management sistemico vitale* (Vol. 1). Torino: Giappichelli.

Barile, S., & Saviano, M. (2010). A new perspective of systems complexity in service science. *Impresa, Ambiente, Management, 4*(3), 375–414.

Barile, S., & Saviano, M. (2011). Foundations of systems thinking: The structure-system paradigm. In Various Authors, *Contributions to theoretical and practical advances in management. A viable systems approach (VSA)*. ASVSA, Associazione per la Ricerca sui Sistemi Vitali (pp. 1–24). Avellino: International Printing.

Barile, S., Pels, J., Polese, F., & Saviano, M. (2012). An introduction to the viable systems approach and its contribution to marketing. *Journal of Business Market Management, 5*(2), 54–78.

Barile, S., Saviano, M., & Simone, C. (2015a). Service economy, knowledge, and the need for T-shaped innovators. *World Wide Web, 18*(4), 1177–1197.

Barile, S., Saviano, M., & Caputo, F. (2015b), How are markets changing? The emergence of consumers market systems. In G. Dominici (Ed.), *The 3rd International Symposium Advances in Business Management. "Towards Systemic Approach"* (pp. 203–207). Avellino: Busyness Systems.

Barile, S., Saviano, M., Polese, F., & Caputo, F. (2015c). T-shaped people for addressing the global challenge of sustainability. In E. Gummesson, C. Mele, & F. Polese (Eds.), *Service dominant logic, network and systems theory and service science: Integrating three perspectives for a new service agenda* (pp. 1–18). Napoli: Giannini.

Barile, S., Lusch, R., Reynoso, J., Saviano, M., & Spohrer, J. (2016). Systems, networks, and ecosystems in service research. *Journal of Service Management, 27*(4), 652–674.

Bartels, R. (1976). *The history of marketing thought*. New York: Publishing Horizon.

Beer, S. (1979). *The heart of enterpris*. New York: John Wiley & Sons.

Bettis, R. A., & Prahalad, C. K. (1995). The dominant logic: Retrospective and extension. *Strategic Management Journal, 16*(1), 5–14.

Bohman, J. (1993). *New philosophy of social science: Problems of indeterminacy*. Boston: The MIT Press.

Brown, S. L., & Eisenhardt, K. M. (1997). The art of continuous change: Linking complexity theory and time-paced evolution in relentlessly shifting organizations. *Administrative Science Quarterly, 42*(1), 1–34.

Bryman, A., & Bell, E. (2015). *Business research methods*. Cambridge: Oxford University Press.

Byrne, D. S. (1998). *Complexity theory and the social sciences: An introduction*. London: Psychology Press.

Capra, F. (1985). Criteria of systems thinking. *Futures, 17*(5), 475–478.

Caputo, F. (2016). A focus on company-stakeholder relationships in the light of the stakeholder engagement framework. In D. Vrontis, Y. Weber, & E. Tsoukatos (Eds.), *Innovation, entrepreneurship and digital ecosystems* (pp. 455–470). Cyprus: EuroMed Press.

Caputo, F., & Walletzký, L. (2017). Investigating the users' approach to ICT platforms in the city management. *Systems, 5*(1), 1–15.

Caputo, F., Del Giudice, M., Evangelista, F., & Russo, G. (2016a). Corporate disclosure and intellectual capital. The light side of information asymmetry. *International Journal of Managerial and Financial Accounting, 8*(1), 75–96.

Caputo, F., Evangelista, F., & Russo, G. (2016b). Information sharing and communication strategies: A stakeholder engagement view. In D. Vrontis, Y. Weber, & E. Tsoukatos (Eds.), *Innovation, entrepreneurship and digital ecosystems* (pp. 436–442). Cyprus: EuroMed Press.

Caputo, F., Formisano, V., Buronova, B., & Walletzky, L. (2016c). Beyond the digital ecosystems view: insights from smart communities. In D. Vrontis, Y. Weber, & E. Tsoukatos (Eds.), *Innovation, entrepreneurship and digital ecosystems* (pp. 443–454). Cyprus: EuroMed Press.

Cova, B., & Dalli, D. (2009). Working consumers: The next step in marketing theory? *Marketing Theory, 9*(3), 315–339.

Del Giudice, M., Caputo, F., & Evangelista, F. (2016). How are decision systems changing? The contribution of social media to the management of decisional liquefaction. *Journal of Decision Systems, 25*(3), 214–226.

Di Nauta, P., Merola, B., Caputo, F., & Evangelista, F. (2015). Reflections on the role of university to face the challenges of knowledge society for the local economic development. *Journal of Knowledge Economy*, 1–19.

Dominici, G., Yolles, M., & Caputo, F. (2017). Decoding the dynamics of value cocreation in consumer tribes: An agency theory approach. *Cybernetics and Systems, 48*(2), 84–101.

Doyle, P., & Stern, P. (2006). *Marketing management and strategy*. London: Pearson Education.

Dunfee, T. W., Smith, N. C., & Ross Jr, W. T. (1999). Social contracts and marketing ethics. *The Journal of Marketing, 63*(3), 14–32.

Easton, G. (2002). Marketing: A critical realist approach. *Journal of Business Research, 55*(2), 103–109.
Edvardsson, B., Tronvoll, B., & Gruber, T. (2011a). Expanding understanding of service exchange and value co-creation: A social construction approach. *Journal of the Academy of Marketing Science, 39*(2), 327–339.
Edvardsson, B., Tronvoll, B., & Höykinpuro, R. (2011b). Complex service recovery processes: How to avoid triple deviation. *Managing Service Quality: An International Journal, 21*(4), 331–349.
Espejo, R. (1994). What is systemic thinking? *System Dynamics Review, 10*(2–3), 199–212.
Evangelista, F., Caputo, F., Russo, G., & Buhnova, B. (2016). Voluntary corporate disclosure in the era of social media. In F. Caputo (Ed.), *The 4rd International Symposium Advances in Business Management. "Towards Systemic Approach"* (pp. 124–128). Avellino: Business Systems. E-book series.
Fabrigar, L. R., Wegener, D. T., MacCallum, R. C., & Strahan, E. J. (1999). Evaluating the use of exploratory factor analysis in psychological research. *Psychological Methods, 4*(3), 272.
Ferrell, O. C., & Gresham, L. G. (1985). A contingency framework for understanding ethical decision making in marketing. *The Journal of Marketing, 49*(3), 87–96.
Fodness, D. (2005). Rethinking strategic marketing: Achieving breakthrough results. *Journal of Business Strategy, 26*(3), 20–34.
Folke, C., Hahn, T., Olsson, P., & Norberg, J. (2005). Adaptive governance of social-ecological systems. *Annual Review of Environment and Resources, 30*, 441–473.
Fuat Firat, A., & Shultz, C. J. (1997). From segmentation to fragmentation: Markets and marketing strategy in the postmodern era. *European Journal of Marketing, 31*(3/4), 183–207.
Furrer, O., Liu, B. S. C., & Sudharshan, D. (2000). The relationships between culture and service quality perceptions basis for cross-cultural market segmentation and resource allocation. *Journal of Service Research, 2*(4), 355–371.
Gettier, E. (1963). Is justified true belief knowledge? *Analysis, 23*(6), 121–123.
Giddens, A. (1991). *Modernity and self-identity: Self and society in the late modern age*. Palo Alto: Stanford University Press.
Glasersfeld, E. V. (1996). Introduction: Aspects of constructivism. In C. T. Fosnot (Ed.), *Constructivism: Theory, perspectives, and practice* (pp. 3–7). London: Teachers College Press.
Goldfarb, A., Ho, T. H., Amaldoss, W., Brown, A. L., Chen, Y., Cui, T. H., & Xiao, M. (2012). Behavioral models of managerial decision-making. *Marketing Letters, 23*(2), 405–421.
Golinelli, G. M. (2010). *Viable systems approach (VSA): Governing business dynamics*. Padova: Cedam.
Golinelli, G. M., Barile, S., Saviano, M., & Polese, F. (2012). Perspective shifts in marketing: Toward a paradigm change? *Service Science, 4*(2), 121–134.
Granovetter, M. (2005). The impact of social structure on economic outcomes. *The Journal of Economic Perspectives, 19*(1), 33–50.
Griswold, C. (2001). *Platonic writings/platonic readings*. New York: Penn State Press.
Gummesson, E., & Polese, F. (2009). B2B is not an island! *Journal of Business and Industrial Marketing, 24*(5/6), 337–350.
Hills, G. E., Hultman, C. M., & Miles, M. P. (2008). The evolution and development of entrepreneurial marketing. *Journal of Small Business Management, 46*(1), 99–112.
Hochschild, A. R. (1979). Emotion work, feeling rules, and social structure. *American Journal of Sociology, 85*(3), 551–575.
Holling, C. S. (2001). Understanding the complexity of economic, ecological, and social systems. *Ecosystems, 4*(5), 390–405.
Hunt, S. D. (1994). On rethinking marketing: Our discipline, our practice, our methods. *European Journal of Marketing, 28*(3), 13–25.
Hunt, S. D. (2002). *Foundations of marketing theory: Toward a general theory of marketing*. Armonk, NY: ME Sharpe.

Inglehart, R. (1997). *Modernization and postmodernization: Cultural, economic, and political change in 43 societies*. Princeton, NJ: Princeton University Press.
Kauffman, S. (2007). Beyond reductionism: Reinventing the sacred. *Zygon®, 42*(4), 903–914.
Kline, S. J. (1995). *Conceptual foundations for multidisciplinary thinking*. Palo Alto: Stanford University Press.
Kohli, A. K., & Jaworski, B. J. (1990). Market orientation: The construct, research propositions, and managerial implications. *The Journal of Marketing, 54*(2), 1–18.
Kumar, V. (2015). Evolution of marketing as a discipline: What has happened and what to look out for. *Journal of Marketing, 79*(1), 1–9.
Lemmink, J. (2005). The need for more multidisciplinary research. *International Journal of Service Industry Management, 16*(1), 7–9.
Letaifa, S. B., Edvardsson, B., & Tronvoll, B. (2016). The role of social platforms in transforming service ecosystems. *Journal of Business Research, 69*(5), 1933–1938.
Lusch, R. F., & Vargo, S. L. (2014). *The service-dominant logic of marketing: Dialog, debate, and directions*. London: Routledge.
Maclaran, P. (Ed.). (2009). *The SAGE handbook of marketing theory*. London: Sage Publication.
Mahoney, M. J. (2004). What is constructivism and why is it growing. *Contemporary Psychology, 49*(3), 360–363.
Mavondo, F., & Farrell, M. (2003). Cultural orientation: Its relationship with market orientation, innovation and organisational performance. *Management Decision, 41*(3), 241–249.
Mele, C., Pels, J., & Polese, F. (2010). A brief review of systems theories and their managerial applications. *Service Science, 2*(1-2), 126–135.
Miller, J. H., & Page, S. E. (2009). *Complex adaptive systems: An introduction to computational models of social life*. New York: Princeton University Press.
Möller, K., & Halinen, A. (2000). Relationship marketing theory: Its roots and direction. *Journal of Marketing Management, 16*(1-3), 29–54.
Morgan, N. A., Clark, B. H., & Gooner, R. (2002). Marketing productivity, marketing audits, and systems for marketing performance assessment: Integrating multiple perspectives. *Journal of Business Research, 55*(5), 363–375.
Moulaert, F. (Ed.). (2013). *The international handbook on social innovation: Collective action, social learning and transdisciplinary research*. London: Edward Elgar Publishing.
Palmer, A., & Ponsonby, S. (2002). The social construction of new marketing paradigms: The influence of personal perspective. *Journal of Marketing Management, 18*(1-2), 173–192.
Patterson, P. G., & Spreng, R. A. (1997). Modelling the relationship between perceived value, satisfaction and repurchase intentions in a business-to-business, services context: An empirical examination. *International Journal of service Industry management, 8*(5), 414–434.
Perez, C. (1983). Structural change and assimilation of new technologies in the economic and social systems. *Futures, 15*(5), 357–375.
Peters, L. D., Pressey, A. D., Vanharanta, M., & Johnston, W. J. (2013). Constructivism and critical realism as alternative approaches to the study of business networks: Convergences and divergences in theory and in research practice. *Industrial Marketing Management, 42*(3), 336–346.
Plouffe, C. R. (2008). Examining "peer-to-peer" (P2P) systems as consumer-to-consumer (C2C) exchange. *European Journal of Marketing, 42*(11/12), 1179–1202.
Polese, F. (2009). The influence of networking culture and social relationships on value creation. *Quaderni di Sinergie, Firms' Management: Processes, Networks and Value, 16*, 193–215.
Pride, W., & Ferrell, O. C. (2016). *Foundations of marketing*. Scarborough: Nelson Education.
Rabinow, P. (1987). *Interpretive social science: A second look*. Berkeley: University of California Press.
Redman, C. L., Grove, J. M., & Kuby, L. H. (2004). Integrating social science into the long-term ecological research (LTER) network: Social dimensions of ecological change and ecological dimensions of social change. *Ecosystems, 7*(2), 161–171.

Roberts, J. H. (2000). Developing new rules for new markets. *Journal of the Academy of Marketing Science, 28*(1), 31–44.
Rubenstein-Montano, B., Liebowitz, J., Buchwalter, J., McCaw, D., Newman, B., Rebeck, K., & Team, T. K. M. M. (2001). A systems thinking framework for knowledge management. *Decision Support Systems, 31*(1), 5–16.
Sarasvathy, S. D. (2001). Causation and effectuation: Toward a theoretical shift from economic inevitability to entrepreneurial contingency. *Academy of Management Review, 26*(2), 243–263.
Saviano, M., Parida, R., Caputo, F., & Datta, S. K. (2014). Health care as a worldwide concern. Insights on the Italian and Indian health care systems and PPPs from a VSA perspective. *EuroMed Journal of Business, 9*(2), 198–220.
Saviano, M., Caputo, F., Formisano, V., & Walletzký, L. (2016). From theory to practice: Applying systems thinking to smart cities. In F. Caputo (Ed.), *The 4rd International Symposium Advances in Business Management. "Towards Systemic Approach"* (pp. 35–40). Avellino: Business Systems. E-book series.
Smith, N. C., Drumwright, M. E., & Gentile, M. C. (2010). The new marketing myopia. *Journal of Public Policy and Marketing, 29*(1), 4–11.
Srivastava, R. K., Shervani, T. A., & Fahey, L. (1999). Marketing, business processes, and shareholder value: An organizationally embedded view of marketing activities and the discipline of marketing. *The Journal of Marketing, 63*(1), 168–179.
Stacey, R. D. (1995). The science of complexity: An alternative perspective for strategic change processes. *Strategic Management Journal, 16*(6), 477–495.
Tadajewski, M. (2004). The philosophy of marketing theory: Historical and future directions. *The Marketing Review, 4*(3), 307–340.
Tadajewski, M., & Jones, D. G. B. (2009). *History of marketing thought*. London: Sage.
Toffler, A., & Alvin, T. (1981). *The third wave* (pp. 32–33). New York: Bantam books.
Tronvoll, B. (2011). Negative emotions and their effect on customer complaint behaviour. *Journal of Service Management, 22*(1), 111–134.
Tronvoll, B. (2012). A dynamic model of customer complaining behaviour from the perspective of service-dominant logic. *European Journal of Marketing, 46*(1/2), 284–305.
Tsai, S. P. (2005). Integrated marketing as management of holistic consumer experience. *Business Horizons, 48*(5), 431–441.
Tushman, M. L., & Romanelli, E. (2008). Organizational evolution. In W. Burke, D. G. Lake, & J. W. Paine (Eds.), *Organization change: A comprehensive reader* (pp. 155–174). New York: John Wiley & Sons.
Vargo, S. L., & Lusch, R. F. (2004). Evolving to a new dominant logic for marketing. *Journal of Marketing, 68*(1), 1–17.
Vargo, S. L., & Lusch, R. F. (2011). It's all B2B... and beyond: Toward a systems perspective of the market. *Industrial Marketing Management, 40*(2), 181–187.
Vargo, S. L., Maglio, P. P., & Akaka, M. A. (2008). On value and value co-creation: A service systems and service logic perspective. *European Management Journal, 26*(3), 145–152.
Verschuren, P. J. (2001). Holism versus reductionism in modern social science research. *Quality and Quantity, 35*(4), 389–405.
Webster, F. E., Jr. (2005). A perspective on the evolution of marketing management. *Journal of Public Policy and Marketing, 24*(1), 121–126.
Webster, F. E., Jr., & Lusch, R. F. (2013). Elevating marketing: Marketing is dead! Long live marketing! *Journal of the Academy of Marketing Science, 41*(4), 389–399.
Werhane, P. H. (2008). Mental models, moral imagination and system thinking in the age of globalization. *Journal of Business Ethics, 78*(3), 463–474.
Wilmott, H. (2001). Breaking the paradigm mentality. In W. O. B. Staff (Ed.), *Organizational studies: Critical perspectives on business and management* (Vol. 1, pp. 625–666). London: Psychology Press.
Wolf, S. (1999). Toward a systemic theory of informed consent in managed care. *Houston Law Review, 35*(5), 1631–1681.

Chapter 2
Successful Value Co-creation Exchanges: A VSA Contribution

Francesco Polese

Abstract According to a growing research community there is the awareness that successful service exchanges are at the base of positive interactions among socio-economic actors. Within this community value co-creation is among the most emerging research focus of many scholars addressing service exchange; nevertheless the concept, indeed intriguing by itself, deserves attention in order to better understand its inner traits. Literature, in this sense, appears to treat value co-creation as a dogma, and more effort can be focused in declining its functioning mechanisms. With this scope this chapter deepens the structural and behavioural enablers of successful value co-creation exchanges, based upon wise resource integration among engaged actors. To reach this scope the manuscript introduces two relevant research streams represented by Service Dominant logic (S-D logic) and the Viable Systems Approach (VSA), both contributing to a better understanding of the investigated phenomenon.

Keywords Value co-creation • Service dominant logic • Viable systems approach • Service exchange

2.1 Introduction

In recent decades, the relevant role of value co-creation concept in resource exchange has been confirmed by literature approaching this subject from different perspectives (Normann and Ramirez 1993; Payne and Holt 2001; Gummesson 2004a, b; Prahalad and Ramaswamy 2004a, b; Grönroos 2008). In particular, within the service community, SD Logic has recently introduced a scientific proposition supporting the diverse and dynamic nature of interactions among different actors—*such as persons, organizations, and firms*—when integrating resources and facilitating service exchange.

F. Polese (✉)
Department of Business Sciences – Management and Innovation Systems, University of Salerno, Fisciano, Italy
e-mail: fpolese@unisa.it

S. Barile et al. (eds.), *Social Dynamics in a Systems Perspective*, New Economic Windows, DOI 10.1007/978-3-319-61967-5_2

Focusing on the analysis of relationships and on the reasons behind the activation of these relationships during the interaction of elements, this work examines the structural and behavioral *enablers of viability* related to the preconditions that create the best possible ecosystem for co-creation. From a behavioral point of view, we argue that strengthening systemic and dynamic traits should be based on actors' appropriate attitudes and consequent dynamics, as well as on effective resource integration and on the non-conflicting (and possibly aligned) goals of survival demonstrated by the involved actors. These actors—with their appropriate attitudes and shared goals—are motivated, and they are most likely willing to develop harmonic interactions by offering and integrating the resources needed for 'viable' service exchanges (Xie et al. 2008; McColl-Kennedy et al. 2013).

Service Dominant (S-D) logic offers interesting insights on this issue because its underlying theoretical framework is based on resource integration as a key driver of successful value co-creation among actors. Moreover, systems theories, and specifically the Viable Systems Approach (VSA), support the process of identifying "successful value co-creation" by locating the meaning of "viable value co-creation" inside the previous concept, deepening the intended meaning. These two research streams, if integrated, could be helpful in the process analyzing this type of relation, which goes beyond the concept of value co-creation and focuses its attention much more on the relation itself rather than the actors.

Therefore, the aim of this work is to interpret the concept of successful value co-creation through the Viable Systems Approach (VSA) framework. The VSA presents a metamodel based upon viable systems generated by interactive actors searching for their own viability and for the viability of the emergent system. Fundamentally, the VSA is a lens through which to observe the complex phenomena that characterize contexts today. Thus, it offers a useful approach to observing service exchange when considering that exchanges are based upon the extant and emerging interactions among many actors, which generate increasing entropy. The VSA, from this perspective, supports the search for viable exchanges when resource integration is fulfilled thanks to the alignment of involved actors' goals and expectations.

Value co-creation is, in fact, based on the interaction between actors who integrate resources to generate benefits; many works in the literature have described value co-creation, but, in this contribution, the VSA perspective on value co-creation is presented through an investigation of successful value co-creation exchanges. This type of relation not only creates benefits but is generated by interactive actors that share the goal of survival, thus keeping the interaction alive and stable and generating the emergence of the system. Successful value co-creation is a viable form of value co-creation that does not have a destination point; it is not a goal but rather a path through continuous research on interactions in order to ensure the system's survival. In other words, value co-creation is not dogmatic. It is an ideal interaction among service actors; this means that, if it is not 'successful,' value co-creation descends from actors that are not able to iterate and promote future positive interactions among themselves.

Successful value co-creation represents effective resource integration between actors and with structural and behavioral enablers of viable behaviors, and for this reason, from a static point of view, it is related to the presence of a predisposition to viability (consonance), whereas from a dynamic point of view, it may be related to the will to survive (resonance). By reinterpreting value co-creation through VSA, the work aims at addressing the existing lack in literature concerning a holistic understanding (mediating between reductionism and holism) of the concept (Wieland et al. 2012). Moreover, the analysis of the key levers enabling value co-creation can lead practitioners and managers to identify new strategies for optimizing resources exchange.

The chapter begins with an overview of the literature on service and management, shifting from the role of the customer to the role of the actor; the focus is on the interactions between the actors and on the concepts of involvement, engagement and self-engagement. The chapter continues with a presentation of the S-D logic and VSA frameworks useful for viability analysis. The core of the chapter, which describes successful value co-creation, discusses the definition of the structural and behavioral enablers of successful value co-creation, enjoying a profitable parallel with viability, among the interactive actors. After the presentation of an example of successful value co-creation in the healthcare system, the conclusions are presented.

2.2 Literature Overview

2.2.1 Advances in Service and Management Research on the Customer's Role

Over the years, business studies have paid significant attention to customer participation and its impact on the service provision process. Many scholars have investigated aspects of critical issues related to customer participation, and related analyses have been performed in the marketing literature (for a complete review, see Mustak et al. 2013). For instance, Lengnick-Hall (1996) identified the following five roles of customers: resource, co-creator, buyer, user, and product, followed by identification of customer-experienced outcomes. However, initial studies in the field of business management performed in-depth analyses of the process of co-production in service management rather than analyses of the broader concept of '*participation*'. In this regard, Toffler (1980) used the term *prosumer* to describe the dual role played by the customer in the delivery process, i.e., as both co-producer and consumer of the service. Gummesson (2007) used this term (i.e., *prosumer*) to highlight how the customer is an integral part of the production system and the progressive change in the logic of service delivery.

2.2.2 *Using the Emerging Concept of "Enabler" to Stimulate Customer Participation*

Previously, Normann (2001) recognized that greater customer participation implies that service organizations must progressively change their role from that of *relievers*, who perform an activity for customers, to that of *enablers,* who put customers in a position to perform tasks on their own by giving them the necessary knowledge and tools. Other scholars have also explored how participation assumes a crucial role in the customer's satisfaction (e.g., by developing the customer's sense of control or providing a highly customized service) and in the service provider's efficiency (e.g., by transferring phases of service to the customer) (Bower et al. 1990; De Santo et al. 2011). Conversely, Dong and Sivakumar (2015) developed a classification of customer involvement based on the participation process (structured or unstructured) and service output (generic or specific output) that highlights how the effects of customer participation on satisfaction and efficiency can differ depending on the type of process or output considered. There are a number of different business policies that a service provider can adopt to encourage its customers to participate more actively (Bitner et al. 1997; Etgar 2008). However, effective customer participation is influenced by many different factors, such as customer loyalty, frequency of service adoption, and positive attitudes toward innovation. However, not all customers are equally motivated to participate; thus, service providers should try to determine which segments of the customer population are more and which are less oriented toward participation. Then, the provider can develop differentiated policies either to facilitate the participation of active customers or to encourage passive customers to participate.

2.2.3 *The Contribution of Service Research and the Concepts of Involvement, Engagement and Self-Engagement*

Service research has also contributed significantly to the understanding of '*involvement*', a concept initially used by Prahalad and Ramaswamy (2000) with reference to the co-opting customer in the value-creating process, implicitly underlining the active role of the provider in promoting the customer's participation. Lusch et al. (2007) use the term involvement, attributing to it a meaning that stresses the pro-active role of the provider in facilitating the active participation of the customer. Bettencourt (1997) uses the term involvement in a way that places greater emphasis on the role of the client in the process of co-creation, putting the client on the same level as the provider and highlighting their reciprocal involvement. Payne et al. (2008) use the term involvement to analyze the role of the customer as a co-creator of value, placing him/her at the same level of importance as the provider. They also use the term 'emotional 'engagement'; Finally, Etgar (2008) develops a model of consumer involvement in co-production.

Building on mainstream service research, however, the term '*engagement*' has been increasingly discussed. Kumar et al. (2010) propose several components of a customer's engagement value and underline the term's reciprocity in the relationship between customer and service provider. Sweeney et al. (2015) use the term 'engagement' to analyze the holistic nature of the customer's role in the customer experience, and Sweeney et al. (2015) do the same in defining the customer's effort in value co-creating activities (EVCA). Recently, the term 'engagement' has been adopted by (S-D) logic researchers to define the active, equal and reciprocal participation of both the customer and the provider in the co-generation of value. In this sense, Shaw et al. (2011) analyze the role of the customer by applying S-D logic in the context of tourism management.

Another aspect of engagement has been deployed to deepen the concept of '*self-engagement*.' This approach has been led by the in-depth analyses conducted by psychology scholars. Britt (1999), in the context of employee engagement, emphasizes the relation between individuals' performance and active participation. According to this relation, self-engagement is defined as "individuals feeling a sense of responsibility for and commitment to a performance domain so that performance matters to the individual" (Britt et al. 2005, p. 1476). With respect to other models of participation, this concept highlights the subject's sense of responsibility, creating a model in which the intensity of the customer's participation is beyond the intensity of normal engagement, particularly in terms of the aspect of dynamism, which determines the development of positive and harmonious interactions with the service provider. Recently, the term actor was introduced to describe the variegated and interchangeable roles of all participants in service exchanges (Gummesson and Polese 2009; Wieland et al. 2012).

2.3 The Theoretical Framework of Service Exchanges: Contributions from SD-Logic and VSA

As presented in the introduction, the S-D logic stream offers interesting support by describing the concept of enablers in particular. It also includes the theoretical framework of resource integration as a key driver of successful value co-creation among actors. The elements discussed below are useful in reorganizing the roots of this approach. Moreover, systems theories, and specifically the Viable Systems Approach (vSa), support the process of identifying "successful value co-creation" by a profitable parallelism with the concept of "viable value co-creation". The integration and the mutual support of the two research streams could be helpful in going beyond the concept of value co-creation so as to focus attention more on the relations than on the actors.

2.3.1 Relevant Elements in S-D Logic

S-D logic is a theoretical proposal originally focused on marketing but later generalized to the functioning of markets, to general management and all its subdisciplines and to economics and society in general. It highlights a paradigm shift away from the goods-dominant (G-D) logic, which has lingered in mainstream management thinking since the advent of the industrial era. For marketing, Vargo and Lusch offer a new perspective by introducing the dominance of services over products and goods, thus adapting to today's competitive service economy (Levitt 1981). S-D logic is based on eleven foundational premises (FPs) and five axioms (Vargo and Lusch 2016). According to these, service should be understood as an application of competences through activities, processes and performances designed to produce benefits for suppliers and customers and for all third parties that are directly or indirectly involved in a network of relationships (Vargo and Lusch 2008).

The processes of value creation, according to S-D logic (Vargo and Lusch 2004), suggest a change in the roles and dimensions of relevance of engaged actors starting from, but not limited to, customers. The concept of the customer was born in the business environment, but a shift from customer to actor is now happening. In this study in particular, it is necessary to focus on the actors because the concept of *enabler of viability* is always valid between actors.

Any case could effectively illustrate the concept of resource integration by presenting the relationship between company and customer, but at the same time, the shift in the role of the actors is useful to explain the concept of value co-creation. The modern meaning of service may be associated with a form of co-creation of value that involves various parties in the service exchange (Prahalad and Ramaswamy 2000, 2004a, b; Mele and Polese 2011). In this service exchange, the actors are interacting with other actors and with the environment; no single actor or provider can realize a complete co-creative experience by itself (Gummesson 2004a, b; Gummesson and Polese 2009). Actors do not obtain value directly from the product itself but rather from its use, processing or consumption and by comparing it with other entities interested or involved in the building process (Katzan 2008). The value of a product is thus derived from the benefits obtainable from the underlying service and from the processes of co-production, co-design, and co-marketing, which involve multiple contributions from different entities (including end users) thanks to the sharing of information, resources, skills, needs and risks (Botti et al. 2017). In the SD-Logic approach, value is therefore determined by the consumer at the time of purchase, through a personal 'consumption' process favored by constant interaction with the other parts of the service system in which it operates (Vargo and Lusch 2008). This process helps us to understand that the actor becomes a real co-creator of value, and consequently, the first actor/value provider (such as a company) is observed only as an integrator (and manager) of the resources necessary for co-creation exchanges.

The S-D logic literature (Lusch et al. 2007; Mustak et al. 2016) highlights the significant role played by the involvement of the customer/actor with regard to value co-creation. Vargo and Lusch (2004) give the term 'involvement' the same meaning by analyzing the service-centered view of exchange and subsequently indicating participation in co-production. Involvement, engagement and self-engagement are often employed as synonyms. Within the terminology adopted by S-D logic, Koskela-Huotari et al. (2013), for example, highlighted and clarified the confusion related to 'co-concepts'. The same emphasis and clarification would be appropriate for the terms used to indicate actor/customer participation, which constitutes a macro container in which the different terms are inserted (e.g., involvement, engagement or self-engagement) and that highlights the different levels of effort of the customer as a resource integrator in the context of the co-creation of value (Lusch and Vargo 2006).

The above discussion explains the root useful to uphold the basic concepts of the enablers of viability; in particular, this root provides a system of meaning to the concept of enabler itself. The activation of a process inside the element could derive from external or internal inputs that generate the need for resource integration.

2.3.2 Systems Theories and *vSA*

The viable systems approach (*vSA*) is a systems theory rooted in systems thinking, and it may well be intended as a meta-model, an interpretive key that is useful for the observation of complex phenomena. It is used for the analysis of relationships among socio-economic entities and in searches for viable interacting conditions (Golinelli 2000; Barile 2008). Among the pillars of system theories are the concepts of open and closed systems (von Bertalanffy 1972), socio-technical systems (Emery and Trist 1960), the law of requisite variety (Ashby 1958), viable systems modeling (Beer 1972) and systems dynamics.

The *vSA* (Golinelli 2000; Barile 2008) proposes a shift in focus 'from the parts to the whole' (Capra 1997), assuming that the organizations could be intended as systems wherein each system is represented by a set of interacting elements (sub-systems) with the same aim: to survive in the long run. Working together, every element promotes its own interests, integrating resources and optimizing competences.

vSA is linked with network analysis; it is based on general system theory and more specifically on social analysis, which interprets business behavior within a dense pattern of interactions. Any socio-economic actor is a viable system in itself and is part of a context of other viable systems and single components. The systemic understanding of actors—and of the relevance of social and business relationships in local environments—affects the actors' behavior, survival capacity and future evolution (Barile 2008). As a systemic theory, *vSA* offers a methodology for interpreting and managing the contemporary business arena. It seems useful for a better understanding of complex dynamics, such as those in which actors are engaged and iteratively interact with each other, as in value co-creation exchanges in general and, of course, in healthcare. The *vSA* proposal, therefore, tries to valorize both holism and reductionism (von Bertalanffy 1956), rejecting the idea that a certain phenomenon can be understood exclusively through an analytical, reductionist approach. Analyzing complex, emerging phenomena exposed to external influences calls for more than an analysis of the interaction among a few components. In elevating and broadening our perspective, therefore, there is a shift in attention from the parts to the whole, with the observed reality being perceived as an integrated and interacting unity of phenomena and the properties of the individual parts becoming less distinct. This shift in perspective, however, implies that the relationships between the parts and the events they produce assume greater significance (Luhmann 1990), which is particularly relevant for a better understanding of co-creation exchanges in service ecosystems. A holistic approach is therefore needed to better understand complex phenomena (Polese et al. 2016).

Identifiable as a core topic in scientific and epistemological debate today, and pervading all disciplines due to its cross-cutting content, complexity, on one hand, reflects the attention that researchers of different disciplines are paying to this issue and, on the other, underlies the contradictions and incongruities related to the different methodological approaches of various disciplines (Barile 2008). According to the *vSA*, complexity is a relative concept that is never absolute, and it can only be assessed in terms of specific contexts of reference, where it refers to a particular combination of multiplicities and autonomies that defy explanation. In the systems approach, the decision-maker needs to distinguish among (i) 'variety' (which refers to possible variants of a phenomenon that might present to the observer at a given time); (ii) 'variability' (which refers to observed changes in variety over time); and (iii) 'indeterminacy' (which refers to whether it is possible to fully understand a given phenomenon) (Barile and Polese 2010a). By applying such a personal interpretive scheme, the decision-maker can begin to better understand the observed complexity and can achieve viability. To ensure viability, systems have to analyze external changes and the other actors' behavior. They then have to adapt in a manner analogous to Darwin's theory of the adaptive capacity of organisms for survival. According to *vSA*, actors are thus able to compete and survive in a particular context if they engage in continuous dynamic processes of adaptation, transformation, restructuring, and business 're-thinking' (Barile 2008; Barile et al. 2012a, b).

In the attempt to be holistic and to survey the whole, and in the attempt to be reductionist by focusing on the characteristics, parts and single components of every system, *vSA* introduces the two concepts of structure and system. Structure refers to what is static and characterizes a reductionist view of the observed reality, focusing on its components and relations and on how the observed phenomena are constituted. System refers to the dynamics of evolution, within a holistic view of the observed reality, and in this way enables the interpretation of interactions. It focuses on the behavior of the observed actors. In this dynamic interaction, the *vSA* contributes to the design and management of positive interactions among actors. What are the key elements of positive interactions between producers (with their offers) and customers (with their needs displayed in their choices)? These elements are not predetermined, as they are characterized by every customer, and they can sometimes be related to customers' communities or aggregations (Barile and Polese 2010a; Barile et al. 2013). They can be subjectively shaped and, most of all, seem to be strictly personal and context-dependent.

The above is what *vSA* suggests for introducing business behavior in search of consonant and resonant interactions among systemic actors, detailed below. Another key theoretical construct proposed by *vSA* is represented by actors' search for harmonic behavior through consonance and resonance. According to *vSA*, the term 'consonance' refers to the potential compatibility between the elements of a system; it thus refers to a static vision of a potential harmonious relationship. Resonance refers to the effective and concrete harmonic behavior resulting from interactions of actors in the service exchange. Consonance is structural and relational; resonance is systemic and interactional (Barile 2008).

2.4 Defining the Enablers of Viability

In brief, the levels of participation analyzed above highlight different grades of effort/intensity of interactions as well as different actor/customer roles in the management of those interactions. These grades are highlighted as the following initiatives:

- In the involvement initiative, participation is assumed by the service provider, and the customer plays a role that involves less effort/intensity.
- In the engagement initiative, participation is assumed both by the customer and by the service provider, with equal levels of effort/intensity.
- In the self-engagement initiative, participation is assumed by the customer, whose role involves greater effort/intensity.

Of the previously examined levels of participation, that which most resembles the concept of "structural and behavioral enablers of viability" is self-engagement.

As assumed previously, the concept of viability, which stems from the *vSA*, is the expression of the will to survive in a complex environment and, naturally, exists within each actor who is busy integrating his resources in the environment. Sometimes, it can be difficult to create the conditions that support the emergence of viability or to find enablers of viability and, for this reason, it is useful to analyze viability under the lens of *vSA*, particularly using the consonance and resonance constructs (Barile 2008). The consonance of elements/actors/systems with other elements/actors/systems corresponds to the structural compatibility and the exchange of interests at an exact moment and in a specific situation. Going beyond static compatibility, resonance emerges when the element/actor/system shows a will to survive in a relationship with other actors. In this way, dynamic interactions between actors take place, as they work together and demonstrate evidence of the same aim (survival). This demonstration is behavioral and is related to the ability to establish efficient relationships with the surrounding contexts by exchanging and integrating resources. We believe that these enablers fundamentally can be drawn as the basic elements supporting successful value co-creation.

2.5 Successful Value Co-creation Enablers': The Structural Features

The theoretical constructs of structure and system in *vSA* can be helpful in describing the enablers of successful value co-creation. The enablers of viability, and thus of successful value co-creation, have structural and behavioral dimensions that emerge from the interaction between the actors and from the exploration of VSA dichotomies related to consonance-resonance and structure-system. Structural features, in fact, concern the prerequisites to establish a common structural basis for exchanges (consonance) in terms of shared capabilities and background, whereas behavioral features pertain to the key levers fostering the emergence of synergic interactions at a systemic level (resonance) in terms of shared purposes.

The enabler reveals itself if structural features exist and if behavioral conditions are active. In any case, the activation of the structural or behavioral features can be causal or random. The structural features are necessary for—and the causal activation comes from—the will to interact, which should create a better opportunity for resonance between the actors. The second type of activation—the random type—could be an effect of different causal interactions or of randomness.

Enablers have different structural features, which include organizational elements, protocols and procedures, tangible and intangible resources, specialized knowledge, physical environments, and many other features. With different qualification, each one is also present in the human mind. It is not possible to analyze all of them in depth here, but we find it interesting to focus on some of them in groups, as detailed below. The groups are service design, resource setting and relational networks.

2.5.1 Service Design

Each actor needs to design itself and modify, day by day, the features necessary to stimulate compatibility with other actors. Thus, service design is the fundamental relevant feature that is useful for describing the meaning of the value that the enabler must provide. Each enabler contributes to defining the actor's features and, through service design, it is possible to develop the necessary conditions to support the consonance of the actor. The enabler's service design, therefore, involves the need to identify the most suitable provision of competences that are useful to stimulate the random or causal activity of resource integration.

2.5.2 Resource Setting

From a structural point of view, resource setting is a relevant feature that determines the nature of the enabler because it is useful to both the internal resource organization of the actor and to the organization of the system of eventual connection with the external forces, i.e., other actors/systems. Resource setting, of course, is strongly affected by the physical environment in which service is experienced because empowerment is promoted and preserved through the development of synergy between the physical environment and the single actor. The socio-economic context may contribute to facilitating or hindering conditions (Chandler and Vargo 2011) that lead to the development of the actor's potential (e.g., awareness of their rights and expectations, self-determination and sense of responsibility) and, consequently, facilitate or hinder the integration of actor resources. As highlighted in mainstream S-D logic, furthermore, the actor, through participation in the context, may broaden and transform his/her potential and resources into competences that can be positively activated while resource integration takes place (Akaka et al. 2013) and ensure that the potential compatibility among the actors who operate in the context, that is, their concordance (Wieland et al. 2012), gives rise to harmonious interactions. These harmonious interactions summarize the resonance of goals that characterize operations (Wieland et al. 2012), which is consistent with the principles of the *vSa*.

2.5.3 Relational Network

A crucial structural precondition is based upon the correct relational network, represented by the definition of smart communication pathways between engaged actors (Maglio et al. 2009). The enablers' activity is, in fact, closely related to the creation of communicative channels that facilitate the actor's participation within the networks. The communicative environment is related to both individual and organizational communication. Individual communication refers to the quality of the communication that takes place between the actors during the process of resource integration. Several factors—namely consonance conditions as highlighted by the *vSa*—influence the enablers of viability, allow structural compatibility among involved actors and are potential conditions for resource integration. A complete analysis, as described above, allows understanding of the mechanisms that directly promote effective value co-creation.

Certainly, the relational network is at the base of resource integration, and the structural features of the enablers of the relational network complete the set of features useful to activate causal or random enablers. This type of relations goes beyond the concept of value co-creation, focusing much more attention on the relations and not on the actors; for this reason, successful value co-creation is not a final destination but rather a path toward survival.

2.6 Defining the Behavioral Enablers of Successful Value Co-creation

From a dynamic and systemic point of view, the behavioral enablers of viability emerge. From a behavioral point of view, therefore, we are looking for the determinants of successful value co-creation, in an attempt to answer the following question: what are the behavioral enablers that allow successful value co-creation?

We have already mentioned that several structural conditions should occur to create the key preconditions supporting successful value co-creation. Valorizing an intriguing parallel with systems theories and the structure/system dichotomy proposed by the *vSA*, however, we may assume that all these preconditions are static and describe a potential not yet deployed. Only when actors start to activate relationships with consonant structural traits can resonant interactions arise in the service exchange, showing harmonic traits and the fulfillment of satisfaction, efficacy and efficiency within the analyzed service ecosystem. Hence, there are various determinants of effective value co-creation, and we choose to group them, as detailed below, as service emergence, resource integration, and interactive exchange.

2.6.1 *Service Emergence*

The actors' behavior is capable of creating, or not creating, harmonic interactions that represent service exchanges. For instance, actors influence the context through the activation of competences capable of promoting cognitive processes that allow them to understand perceived reality through experience (defined as processes of sensemaking by Weick 1995) and through a new cultural approach and personal attitudes; they develop interactions resonant with the other actors and their modes of action, allowing the emergence of service. In any case, knowledge development and improvement are at the base of service emergence and, for this reason, exist between the behavioral enablers for effective value co-creation.

2.6.2 *Resource Integration*

At a structural level, the engagement of the actors in value co-creation is relevant; these actors possess the resources necessary for the designed service. These actors, therefore, are the key players in the dynamics that occur when each actor interacts with the other, creating effective resource integration and, ultimately, resonant behavior for the benefit of the overall ecosystem when goal sharing takes place definitively. One of the fundamental enablers is, hence, related to how the patient performs his/her role as resource integrator in the process of value co-creation.

Resonant and harmonic interactions, in fact, ought not to be taken for granted. As value co-creation, and ultimately resonance, are related to wise resource integration for the actors' benefit and for the fulfillment of their needs, this resource integration process is performed when actors simultaneously and easily release their possessed resource (needed by other actors) and appreciate the received resource (expected in the service exchange). This process is simplified by the actors' behavior and generous attitudes, which ease resource exchange. This situation shows that each relevant actor has what he/she needs and thus fulfills his/her expectations. However, the key point is related to the fact that actors integrate resources and co-create value according to specific goals and purposes.

Generally, when values and resources do not conflict and are aligned with all the other actors' goals and purposes, the value-creation setting is most profitable and effective resource integration takes place. It is in these situations that each actor experiences the rewards and gratification of expectations and will most likely release his/her resource, which is crucial for the other actors' benefit; this is when resonance conditions occur.

2.6.3 Interactional Exchanges

The stimulus of successful value co-creation considers the active role of the interacting actors not only for a single relation but also in a path of relations, considering the medium- and long run path toward the survival of each actor. The interactional exchange is an enabler that comes directly from the actor's capability of being involved, engaged and self-engaged. To permit the interaction, the actors must be aligned in their goals and capability of interacting. For this reason, communication is undoubtedly important. If the actor encounters behavioral enablers in its interactional exchanges, it is tuned to the path toward successful value co-creation because it has the right degree of openness of the system in actor-to-actor interaction. A more in-depth analysis of the effects generated by successful value co-creation on the ecosystem highlights a series of further enablers that are indirect, although important, and connected to value creation. Knowledge-sharing supports information diffusion, awareness and alignment, all of which contribute to better service provision (Payne et al. 2008).

2.7 Discussion

We assume that all enablers of the viability of actors should be met simultaneously, or one or more actors may not adequately fulfill their role of resource integrator and may not make their support available in value co-creation exchanges. Success, however, requires that expectations be mutually consistent and not in opposition, thus allowing a consonant reciprocal alignment in the process of co-creation of

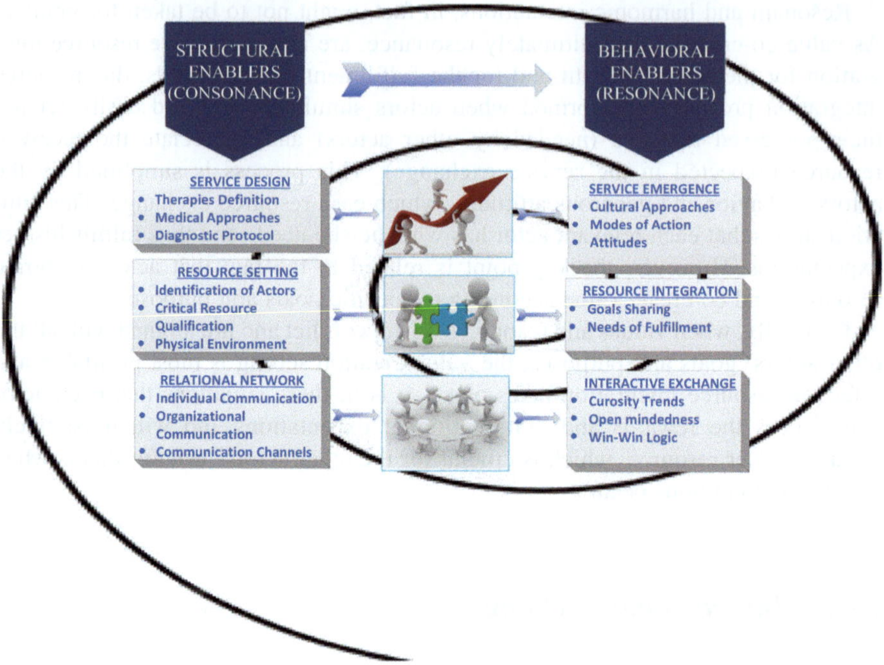

Fig. 2.1 An example of structural and behavioral enablers in the healthcare system

value as a prerequisite for harmonic interactions that characterize resonance. Therefore, decision-makers' actions to promote successful value co-creation should be coherent with the intrinsic enablers of viable behavior.

Within this interpretative framework, the determinant of successful value co-creation represents the dynamic aspects of structural preconditions as described above. With this aim, the service design finds its correspondence in the emergence of service; the resource setting faces the related resource integration; and the relational network calls for interactional exchanges, as graphically represented by Fig. 2.1. To simplify, Fig. 2.1 shows an example of the structural and behavioral enablers that stimulate successful value co-creation in the healthcare system.

The correspondence and integration between structural and behavioral enablers of viability draws a parallel with the structure/system concept (frameworked within the VSA); it reflects an iterative process of self-development and collective-development in all organizations, in which

1. Any specific configuration of elements (including service design, resource setting and relational networks) can allow emerging system interactions between involved actors (and resource owners) to reach common goals by activating static relationships with each other;

2. Several dynamic situations may occur, according to different actors' needs, that match various pathways, as detailed above; all engaged actors should contribute to successful value co-creation if involved;
3. These interactions lead to new configurations and to the use of available resources, and they affect the performances of interacting systems.

This process is iterative. Recent advances in *vSA* studies help us to define and explain this cyclical process by valorizing some of its foundational concepts (FC) (Barile and Polese 2010b; Polese 2013).

According to the structure/system dichotomy (FC6), we assume that the set of structural components (hospital wards, clinics, doctors, nurses, etc.) provides assistance to patients through a virtuous interaction among endogenous components within the service ecosystem. In that regard, system equifinality refers to various system elements capable of achieving the same final state, starting from different initial conditions and passing through different evolutionary paths (Doty et al. 1993; Golinelli 2010).

The context analysis highlights consonant relationships among interested actors operating in the same service ecosystem that are designed to be resonant when harmonious interactions become effective (FC7); in healthcare, this calls for the evaluation of patients' experiences as a form of proper feedback on the participation of each actor in a service exchange; this information is helpful to re-plan future actions.

Moreover, system viability (FC8) relates to the actors' ability to sustain their own activities over time. Viability, therefore, appears to be the final result of all healthcare service operations, thanks to the valorization of contributions received from other involved actors.

System adaptation (FC9) addresses the iterative alignment among critical available resources and modes of action; such alignment is able to fit the healthcare offer to patients' expectations and needs. This leads to an adaptive impulse of healthcare professionals who are in search of positive interactions with all other actors. The system's dynamics, looking for higher levels of interconnection between actors sharing the resources they possess, valorize goal sharing and win-win logic within the healthcare service ecosystem.

2.8 Conclusions and Final Remarks

This chapter describes the concept of successful value co-creation using the frameworks offered by S-D logic and the *vSA*. Fundamentally, the chapter has highlighted that value co-creation cannot be taken for granted; different levels of value co-creation characterize interaction among actors and describe more or less gratifying service exchanges in each involved actor's perspective.

Actors engage among themselves in attempts to fulfill their expectations, and only when they behave in harmony do they realize gratifying service exchanges. In this sense, we have proposed structural preconditions that can, in perspective, bring positive interactions, and we have identified several elements that have the potential to contribute to positive interactions. On the other hand, several enablers have been identified due to their close relation to effective positive interactions among actors.

The reinterpretation of value co-creation through VSA lens involves both theoretical and practical implications. Concerning the first point of view, the identification of the drivers encouraging value co-creation processes can lead researchers to a better understanding of such a complex concept, whose key dimensions, antecedents and consequences are until now relatively unexplored (Yi and Gong 2013; McColl-Kennedy et al. 2013).

From a managerial standpoint, the recognition of the key enablers for successful value co-creation can aid decision- makers in undertaking more effective choices (Barile and Saviano 2010) and in elaborating more adequate strategies for stimulating actor's engagement and for establishing resonant relationships. In addition, the identification of the different structural and systemic prerequisites for co-creation could increase manager's awareness of the whole process and could foster the assessment of the kind of resources exchanged and of user's relational degree at each stage. The holistic evaluation of service exchanges and of eventual problems indering service provision and co-creation can lead to the improvement of service design, offering and delivery (Polese et al. 2016; Ciasullo et al. 2017).

Future studies may support this theoretical proposition by confirming the structural preconditions and effective enablers in specific experimental contexts. As seen in Fig. 2.1, some ideas about the healthcare system are presented, hypothesizing the enablers (in the relative context) and the system of input that causes the actors to generate successful value co-creation. It is not always simple to recognize the structural and behavioral features that create the conditions of viable interactions. In the future, practitioners and scholars may benefit from the theoretical integration of vSA and SD Logic frameworks because both address the emerging contexts and dynamic interactions of actors in the service exchange.

References

Akaka, M. A., Vargo, S. L., & Lusch, R. F. (2013). The complexity of context: A service ecosystems approach for international marketing. *Journal of Marketing Research, 21*(4), 1–20.

Ashby, H. R. (1958). General systems theory as a new discipline. *General Systems (Yearbook of the Society for the Advancement of General Systems Theory), 3*, 1–6.

Barile, S. (2008). *L'impresa come sistema. Contributi sull'approccio sistemico vitale*. Torino: Giappichelli.

Barile, S., Carrubbo, L., Iandolo, F., & Caputo, F. (2013). From 'EGO' to 'ECO' in B2B relationships. *Journal of Business Market and Management, 6*(4), 228–253.

Barile, S., Pels, J., Polese, F., & Saviano, M. (2012a). An introduction to the viable systems approach and its contribution to marketing. *Journal of Business Marketing Management, 2*(2), 54–78.

Barile, S., Saviano, M., Polese, F., & Di Nauta, P. (2012b). Reflections on service systems boundaries: A viable systems perspective. The case of the London Borough of Sutton. *European Management Journal, 30*, 451–465.

Barile, S., & Polese, F. (2010a). Linking the viable system and many-to-many network approaches to service-dominant logic and service science. *International Journal of Quality and Service Science, 2*(1), 23–42.

Barile, S., & Polese, F. (2010b). Smart service systems and viable service systems: Applying systems theory to service science. *Service Science, 2*(1–2), 21–40.

Barile, S., & Saviano, M. (2010). A new perspective of systems complexity in service science. A new perspective of systems complexity in service science. *Impresa, Ambiente, Management, 4*(3), 375–414.

Beer, S. (1972). *Brain of the firm*. London: The Penguin Press.

Bettencourt, L. A. (1997). Customer voluntary performance: Customers as partners in service delivery. *Journal of Retailing, 73*(3), 383–406.

Bitner, J. M., Faranda, W. T., Hubbert, A. R., & Zeithaml, V. A. (1997). Customer contributions and roles in service delivery. *International Journal of Service Industry Management, 8*(3), 193–205.

Botti, A., Grimaldi, M., Tommasetti, A., Troisi, O., & Vesci, M. (2017). Modeling and measuring the consumer activities associated with value cocreation: An exploratory test in the context of education. *Service Science, 9*(1), 63–73.

Bower, M. R., Martin, C. L., & Luker, A. (1990). Trading places: Employees as customers, customers as employees. *Journal of Services Marketing, 4*(2), 55–69.

Britt, T. W. (1999). Engaging the self in the field: Testing the triangle model of responsibility. *Personality and Social Psychology Bulletin, 25*(6), 698–708.

Britt, T. W., Castro, C. A., & Adler, A. B. (2005). Self-engagement, stressors, and health: A longitudinal study. *Personality and Social Psychology Bulletin, 31*(11), 1475–1486.

Capra, F. (1997). *The web of life*. New York: Doubleday-Anchor Book.

Chandler, J. D., & Vargo, S. L. (2011). Contextualization and value-in-context: How context frames exchange. *Marketing Theory, 11*(1), 35–49.

Ciasullo, M. V., Palumbo, R., & Troisi, O. (2017). Reading public service co-production through the lenses of requisite variety. *International Journal of Business and Management, 12*(2), 1–13.

De Santo, M., Pietrosanto, A., Napoletano, P., & Carrubbo, L. (2011). Knowledge based service systems. In E. Gummesson, C. Mele, & F. Polese (Eds.), *System theory and service science: Integrating three perspectives in a new service agenda*. Napoli: Giannini.

Dong, B., & Sivakumar, K. (2015). A process-output classification for customer participation in services. *Journal of Service Management, 26*(5), 726–750.

Doty, D. H., Glick, W. H., & Huber, G. P. (1993). Fit, equifinality, and organizational effectiveness – A test of two configurational theories. *The Academy of Management Journal, 36*(6), 1196–1250.

Emery, F. E., & Trist, E. L. (1960). Socio-technical systems. In C. W. Churchman & M. Verhurst (Eds.), *Management sciences models and techniques* (Vol. 2). London: Pergamon Press.

Etgar, M. (2008). A descriptive model of the consumer co-production process. *Journal of the Academy of Marketing Science, 36*(1), 97–108.

Golinelli, G. M. (2000). *L'approccio sistemico algoverno dell'impresa. L'impresa sistema vitale* (1nd ed.). Padova: CEDAM.

Golinelli, G. M. (2010). *Viable systems approach (vSA): Governing business dynamics*. Padova: CEDAM.

Grönroos, C. (2008). Service logic revisited: Who creates value? And who co-creates? *European Business Review, 20*(4), 298–314.

Gummesson, E. (2004a). Return on relationships (ROR): The value of relationship marketing and CRM in business-to-business contexts. *Journal of Business and Industrial Marketing, 19*(2), 136–148.

Gummesson, E. (2004b). From one-to-one to many-to-many marketing. In B. Edvardsson, A. Gustafsson, S. Brown, & R. Johnston (Eds.), *Service excellence in management: Interdisciplinary contributions, Proceedings from the QUIS 9 Symposium* (pp. 16–25), Karlstad University, Karlstad.

Gummesson, E. (2007). Exit services marketing – Enter service marketing. *Journal of Customer Behaviour, 6*(2), 113–141.
Gummesson, E., & Polese, F. (2009). B2B is not an island! *Journal of Business and Industrial Marketing, 24*(5/6), 337–350.
Katzan, H. (2008). Foundations of service science concepts and facilities. *Journal of Service Science, 1*(1), 1–22.
Koskela-Huotari, K., Friedrich, P., & Isomursu, M. (2013). Jungle of "co". In E. Gummesson, C. Mele, & F. Polese (Eds.), *System theory and service science: Integrating three perspectives in a new service agenda*. Napoli: Giannini Editore.
Kumar, V., Aksoy, L., Donkers, B., Venkatesan, R., Wiesel, T., & Tillmanns, S. (2010). Undervalued or overvalued customers: Capturing total customer engagement value. *Journal of Service Research, 13*(3), 297–310.
Lengnick-Hall, C. A. (1996). Customer contributions to quality: A different view of the customer-oriented firm. *Academy of Management Review, 21*(3), 791–824.
Levitt, T. (1981). Marketing intangible products and products intangibles. *Harvard Business Review, 59*, 94–102.
Luhmann, N. (1990). *Essays on self-reference*. New York: Columbia University Press.
Lusch, R. F., & Vargo, S. L. (2006). Service-dominant logic: Reactions, reflections and refinements. *Marketing Theory, 6*(3), 281–288.
Lusch, R. F., Vargo, S. L., & O'Brien, M. (2007). Competing through service: Insights from service-dominant logic. *Journal of Retailing, 83*(1), 5–18.
Maglio, P. P., Vargo, S. L., Caswell, N., & Spohrer, J. (2009). The service system is the basic abstraction of service science. *Information Systems and e-business Management, 7*(4), 395–406.
McColl-Kennedy, J. R., Vargo, S. L., Dagger, T. S., Sweeney, J. C., & Kasteren, Y. (2013). Health care customer value cocreation practice styles. *Journal of Service Research, 15*(4), 370–389.
Mele, C., & Polese, F. (2011). Key dimensions of service systems in value-creating networks. In B. Hefley, W. Murphy, H. Demirkan, J. C. Spohrer, & V. Krishna (Eds.), *The science of service systems* (pp. 37–59). Boston, MA: Springer US.
Mustak, M., Jaakkola, E., & Halinen, A. (2013). Customer participation and value creation: A systematic review and research implications. *Managing Service Quality, 23*(4), 341–359.
Mustak, M., Jaakkola, E., Halinen, A., & Kaartemo, V. (2016). Customer participation management: Developing a comprehensive framework and a research agenda. *Journal of Service Management, 27*(3), 250–275.
Normann, R. (2001). *Reframing business: When the map changes the landscape*. Chichester: Wiley.
Normann, R., & Ramirez, R. (1993). From value chain to value constellation. *Harvard Business Review, 71*(4), 65–77.
Payne, A., & Holt, S. (2001). Diagnosing customer value: Integrating the value process and relationship marketing. *British Journal of Management, 12*(2), 159–182.
Payne, A. F., Storbacka, K., & Frow, P. (2008). Managing the co-creation of value. *Journal of the Academy of Marketing Science, 36*(1), 83–96.
Polese, F. (2013). *Management sanitario in ottica sistemico vitale*. Torino: Giappichelli Editore.
Polese, F., Tommasetti, A., Vesci, M., Carrubbo, L., & Troisi, O. (2016). Decision-making in smart service systems: A viable systems approach contribution to service science advances. In *International Conference on Exploring Services Science* (pp. 3–14). Bucharest, Romania, 25–27 May 2016, Springer International Publishing.
Prahalad, C. K., & Ramaswamy, V. (2000). Co-opting customer competence. *Harvard Business Review, 78*(1), 79–87.
Prahalad, C. K., & Ramaswamy, V. (2004a). Co-creation experiences: The next practice in value creation. *Journal of Interactive Marketing, 18*(3), 5–14.

Prahalad, C. K., & Ramaswamy, V. (2004b). Co-creating unique value with customers. *Strategy and Leadership, 32*(3), 4–9.

Shaw, G., Bailey, A., & Williams, A. (2011). Aspects of service-dominant logic and its implications for tourism management: Examples from the hotel industry. *Tourism Management, 32*(2), 207–214.

Sweeney, C. J., Danaher, T. S., & McColl-Kennedy, J. R. (2015). Customer effort in value cocreation activities. Improving quality of life and behavioral intentions of health care customers. *Journal of Service Research, 18*(3), 318–335.

Toffler, A. (1980). *The third wave*. New York, NY: Bantam Books.

Vargo, S. L., & Lusch, R. F. (2004). Evolving to a new dominant logic for marketing. *Journal of Marketing, 68*(1), 1–17.

Vargo, S. L., & Lusch, R. F. (2008). Service-dominant logic: Continuing the evolution. *Journal of the Academy of Marketing Science, 36*(1), 1–10.

Vargo, S. L., & Lusch, R. F. (2016). Institutions and axioms: An extension and update of service-dominant logic. *Journal of the Academy of Marketing Science, 44*, 5–23.

Von Bertalanffy, L. (1956). General system theory. *General system, 1*, 1–10.

Von Bertalanffy, L. (1972). The history and status of general systems theory. *Academy of Management Journal, 15*(4), 407–426.

Weick, K. E. (1995). *Sensemaking in organizations*. London: SAGE Publications.

Wieland, H., Polese, F., Vargo, S., & Lusch, R. (2012). Toward a service (Eco)systems perspective on value creation. *International Journal of Service Science, Management Engineering, and Technology, 3*(3), 12–25.

Xie, C., Bagozzi, R. P., & Troye, S. V. (2008). Trying to prosume: Toward a theory of consumers as co-creators of value. *Journal of the Academy of Marketing Science, 36*(1), 109–122.

Yi, Y., & Gong, T. (2013). Customer value co-creation behavior: Scale development and validation. *Journal of Business Research, 66*(9), 1279–1284.

Chapter 3
Complexity and Sustainability in Management: Insights from a Systems Perspective

Sergio Barile and Marialuisa Saviano

Abstract The growing inadequacy of traditional 'management toolkits' for dealing with ever more complex phenomena highlights the need for a rethinking of consolidated management approaches: it is necessary a paradigmatic change for setting a pathway of knowledge and taking into account awareness of the limits and, at the same time, capability of overcoming and updating past schema. Acknowledging the unquestionable value of the long tradition of systems thinking contributions to social sciences and business management, the purpose of this work is to highlight the reasons why a systems approach is really needed to better understand business and social dynamics in condition of complexity, how the Viable Systems Approach can support decision making in condition of complexity, and how the discussion of complexity and the survival of viable systems in a complex scenario, implies inevitably, discussion about sustainability.

Keywords Complexity • Decision making • Systems thinking • Viable systems approach • Sustainability

3.1 Introduction: The Need for Paradigmatic Change in the Management and Study of Organizations

At the end of the 1990s, dissatisfaction with consolidated management models and tools on the part of a small group of Italian scholars of the University of Salerno in Italy became evident. The growing inadequacy of traditional 'management toolkits' for dealing with ever more complex phenomena highlighted the need for radical change in business management approaches. The search on the part of scholars for a more robust and well-grounded organizational management approach provoked a radical rethinking of the general schemes on which the study of organizations

S. Barile
Sapienza University of Rome, Rome, Italy
e-mail: sergio.barile@uniroma1.it

M. Saviano (✉)
University of Salerno, Fisciano, Italy
e-mail: msaviano@unisa.it

S. Barile et al. (eds.), *Social Dynamics in a Systems Perspective*, New Economic Windows, DOI 10.1007/978-3-319-61967-5_3

pivots. In particular, in recent decades, numerous and consolidated models in both the scientific and operational fields have been debated in depth and evaluated with respect to their methodological nature (von Bertallanfy 1967, 1968; Ashby 1968; Beer 1972, 1975; Espejo 1989; Golinelli 2000, 2005; Barile 2009a, b, 2013).

Principles understood as universal also in a personal or individual perspective (above all, in terms of categorical values) are still under debate. Elements seemingly connoted by certainty now find themselves flawed with doubts once again. Growing dissatisfaction with respect to the explicative power of deterministic and reductionist approaches has been experienced by decision makers involved in the management of complexity; therefore, a new trend in terms of defining new conceptual paradigms seemed to be necessary (Ceruti 1986; Simon and De Laplace 1995).

The outcome of the need to overcome a reductionist and deterministic view was principally a shift of focus towards the context. Context was no longer the product of the multiple images the eye records but the outcome of a singular perception, significantly influenced by the specific feelings of the observer. In other words, the context arises from behaviours that constantly filter and frame others through their schema and values: “anything said is said by an observer” (Maturana 1970, p. 8).

The focus was on the entity observed as a whole, and on its structural relations and systemic interactions, and the investigational approach discarded a cause-effect view given that systems thinking deals with complexity and decision-making. Fundamental aspects, in fact, suggest clearly distinguishing between problem solving and decision making as two very different conditions of management processes.

With this distinction, a new risk and uncertainty dichotomy was created. According to the studies conducted by Simon, Cyert and March in the early 1950s, two fundamental postulates around the concept of limited rationality in organizational behaviour were underlined (March and Simon 1958; Cyert and March 1963; Simon 1967; Simon 1991):

– Rationality is limited, given our incapability to define optimal solutions by calculation and our inadequacy in interpreting all the variables we face relative to the context we live in. In this sense, reality posits as true an approach that shapes the concept of rationality on the main characteristics of the decision maker and on the context.
– The evolutionary dynamics of the organizational context provoke the need to constantly redefine the set of problems and variables involved, stressing the impossibility to codify a specific and determined set of ever-valid solutions.

Thus, according to Simon, there is evidence that the rational process is based much more on intellective aspects, on specific and even hidden behavioural competencies than on a given step-by-step procedure. The ‘vertical’ progress of specialization had put the focus on producing and exploiting technical and instrumental knowledge, disregarding the relevance of the method, in the meaning from Greek "μέθοδος",[1] i.e. search, investigation, inquiry—we would say 'exploration'—but also the 'way' of searching that has an irreducible subjective nature. On the one hand, there is the relevance of the patrimony of individual skills and intelligence; on the other, the revaluation of the so-called intellective competencies such as intuition, induction, deduction, logic and reasoning. The turning point in the definition of a new paradigm, however, emerges from the work of two other Authors: Kahneman and Tversky (1982). In particular, their *Prospect Theory* introduces elements of subjective evaluation in choices and puts the accent on the perspective that an individual has relative to the context, the problem itself and possible alternatives where the same problem or the same situation can drive to a different set of choices by virtue of the description given of the problem. The Authors call this the "framing effect".

Therefore, the new risk and uncertainty dichotomy emerges in a managerial context where decisions are made under risk conditions because of the turbulence and the instability of inter-systemic interactions, highlighting the necessity of searching appropriate 'ways' to ensure the survival of the systems managed by adopting a perspective—a 'method'—that could support decision makers in making choices.

Acknowledging the unquestionable value of the long tradition of systems thinking contributions to social sciences and business management, the purpose of this work is to highlight the reasons why a systems approach is needed to understand business and social dynamics in condition of complexity, how the Viable Systems Approach can represent a possible 'method' for decision making in condition of complexity, and how the discussion of complexity and the survival of viable systems in a complex scenario, implies inevitably, discussion about sustainability.

3.2 The Long Tradition of Systems Thinking Studies

The new paradigm for setting a pathway of knowledge had to take into account awareness of the limits but at the same time, also had to be capable of overcoming and updating past schema. The question posed was of course how? The answer albeit long neglected, may now seem elementary. It was necessary to refer back to the great scholars and to re-visit the contribution of systems thinking. In order to give the broadest representation of the various systems interpretations of organizations and firms, a long tradition of systems thinking contributions to social sciences and business management has to be mentioned (Buckley 1967, 1968, 2008; Emery 1969).

It is well known, in fact, that the systems theory has been effectively adopted in several disciplinary domains, it being a general perspective capable of grasping the

[1]It is interesting noting that the word 'μέθοδος' is composed by μετα-, which includes the idea of 'pursuing', and ὁδός, that is 'way'.

rules underpinning the functioning of almost any phenomenon of reality. The well-known General Systems Theory (GST) with its wealth of literature starting from Katz and Kahn (1966), von Bertallanfy (1968), Parsons (1971) and Luhmann (1990a, b), to name but a few, is a case in point. Systems thinking has also been adopted in the managerial field as key to managing complexity and in support of decision-making processes (Golinelli 2005; Barile 2009a, b). A long tradition of studies connotes the systemic human tendency for understanding or interpreting the world around us.

In the main, contributions to systems thinking were linked to the structure and operation of viable systems and their relationship with the environment. Biologists began to study living entities, from a dynamic-evolutionary perspective as complex and integrated wholes (Maturana and Varela 1980); ecologists, rejecting the mechanistic view of the universe (Hannan and Freeman 1977), looked at the earth from an integrated holistic perspective; sociologists and psychologists contributed to the enlightening theory of cognitivism (Clark 1993). The origins of a systems discipline, however, have emerged as the result of constant adjustments within different areas of research. In particular, an initial attempt to build a science of structure based on the principles of organization is owed to Bogdanov (1912). However, it was von Bertallanfy (1968), whose research formally pioneered opportunities for debate on a systems approach, to introduce a General System Theory (GST). Later, with the studies of Stafford Beer (1975), the contribution of cybernetics enriched the body of knowledge around systems. For the first time, the adjective 'viable' was coined with the specific aim of describing an organisational structure of any autonomous system capable of reproducing itself. Beer 's viable system model became a reference point in management studies from its introduction and is currently the basis for the Viable Systems Approach (*vSA*) (Beer 1972, 1974; Espejo 1989; Golinelli 2000; Barile 2005).

This evolutionary process now veering towards a systems view of business organizations, in actual fact, reflected a dominant perspective ranging over time from the mechanistic analogy of the firm as a machine (Taylor 1914), an organic (von Bertallanfy 1968), cybernetic (Beer 1972) and autopoietic system (Maturana 1975), to that of a cognitive system (Hinterhuber et al. 1996). The connotation 'viable' represents the apex of such an evolutionary (as opposed to revolutionary) process based on the observation that reality, knowledge, and even firms have to be understood as a unitary whole of internal relations and external interactions.

After initial fervour, however, there was a gradual weakening of interest around the theme (Barile et al. 2016a). Although appreciated as an emerging, new paradigm by Kuhn (1962) by virtue of its contrasting impact to the limits of mechanistic approaches, the systems approach was not acknowledged as offering a substantially full contribution to organizations and their management (Kast and Rosenzweig 1972), and for this reason it was considered too abstract to be a reliable representation of the specifics of given phenomena. It was argued that the systems approach appeared to lack "disciplined generalizations and rigorous deductions" (Rapoport and Horvath 1968: XII) and GST was considered a "third order study" (Philips 1971). In contrast, "Contingency theory" was classified as a "second order" study, focused on more specific characteristics and relationships in social organizations (Lorsch and Lawrence 1970).

This matrix represents the context in which the Viable System Approach (*vSA*) was introduced. By strongly reaffirming the contribution of systems thinking to management, the original group of Salerno University researchers engaged a growing number of scholars in a shared scientific effort targeted at developing a set of interpretation schemes to adopt for interpreting and understanding the essence, form and behaviour of organizations and firms. As a coherent set of interpretations schemes, rooted in systems thinking and built upon an updated version of the Stafford Beer's Viable System Model (1972), developed to study and govern business as well as other social organizations (Golinelli 2000, 2010, 2011; Barile 2000, 2009a, b, 2013; Various Authors 2011; Barile et al. 2012), *vSA* can be considered a methodology. In the management context, despite the widely accepted view of business as a socio-technological open 'system', its implications have rarely been explored in depth and, as a consequence, many relevant systemic characteristics have not been considered in decision-making processes. Hence, the purpose of this work is to underline the necessity of a systems approach to better understand business and social dynamics, highlighting the contribution of *vSA*.

3.2.1 The Contribution of the Viable Systems Approach to Management Studies

The specific aim of *vSA* is to offer a unitary conceptual framework made up of general interpretation schemes developed on general systems theory principles for interpreting business and social phenomena in terms of their basic dynamics. The contribution of *vSA* adds to existing schemes of synthesis, general schemes useful to move among diverse problematic contexts. This cognitive transversality is grounded in general (or universal) rules and principles of systems thinking that characterize the behaviour of any living (or viable) entity. As a very general set of schema, *vSA* can significantly reinforce the knowledge capabilities of decision makers in conditions of complexity.

To comprehend the specific contribution of *vSA*, we set out herewith the main points of Stafford Beer's model (Beer 1985; Espejo 1993) on which *vSA* is based. According to Beer, a viable system is a "System which survives, remains united and is integral, is homeostatically balanced both internally and externally and possesses mechanisms and opportunities for growth and learning, development and adaptation, which allow it to become increasingly effective within its environment" (Beer 1985[2]).

[2]Summary of the definition given in BEER S., *Diagnosing the system for organization*, John Wiley, 1985.

3.2.1.1 At the Roots of the *vSa*: From the Viable System Model to the Viable Systems Approach

The Stafford Beer model, based on the observation of biological and social systems, recovers and gives centrality to the recursive organization of structures, assigning to the various levels of recursion, different degrees of priority (in Beer's view, these are implementation, coordination, control, intelligence and implementation again). The model is functional for any typology of organization (not only firms) that tries to survive by modifying over time its logical and physical structure. Viability is, accordingly, the essential modality capable of characterizing the nature of social entities. It is experienced when, based on the relations among the internal components of the structure and them and selected external elements of other systems, the organization interacts with the context in order to learn, adapt, develop itself and improve its efficacy in relation to the specific aim of survival.

It is clear that the paternity of the model is still Stafford Beer's and conceptual matrix is essentially the same. However, *vSA* introduces some specific innovations, above all in relation to the analysis of decision-making.

The first pertinent difference is identifiable in the concept of viability. In Beer's view, viability is associated with autonomous existence and arises from the process of decision making within the context of systems made up of other viable entities. Focus, however, is on the system whose main goal is survival in terms of 'autonomous existence'. In *vSA*, focus is more on the context of the system while survival is interpreted as existence 'resonant' with the context. Of course, these are the two sides of the same coin, but this slight difference has important implications, especially in an ecological view of viable systems. We discuss this point further in the following section where we highlight the relevance of the shift from an 'ego' to an 'eco' view of survival of viable systems (Barile et al. 2013).

Another element that requires redefining regards the principle of isotropy. According to Beer, there is a fundamental distinction between government and operations, which is at the basis of the management approach. While substantially complying with this view, *vSA* offers a slightly different view that, complying with the above outlined binominal decision-making—problem solving, distinguishes between governance and management. It is not simply a question of terms: governance refers to the set of principles and rules that characterize the action of the governing body (decision maker) of the system, i.e. the decision-making area. Management on the contrary, refers not only to operations but also to managerial decisional activity performed at operative level as problem solving. Thus, by extracting from the whole set of decisions those we call 'managerial' and unifying them with the operational area, we obtain a reformulated concept of isotropy which, also involving studies on the human's brain functionality, underlines the existence of structural components designed to develop deductive processes, others designed to develop inductive processes: a formulation which is no less organicist compared to that proposed by Beer. This view is illustrated in Fig. 3.1.

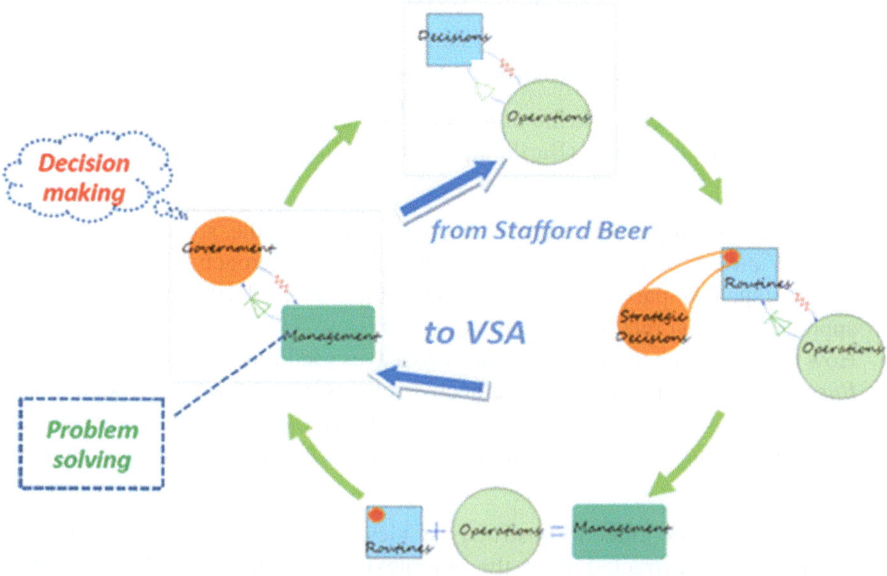

Fig. 3.1 From beer's viable system model to the viable systems approach. Source: Golinelli (2010)

3.2.1.2 Key Concepts and Principles of *vSA*

Another fundamental point concerns the role that the decision maker has in framing and detaching in a constructivist manner, the context from the environment. Decision makers, in fact, focus attention on the external context, which is made up of other systems (so called *suprasystems*). However, the framed context is just an extract of what the decision maker is able to see, according to the schemas and the values he/she has. Through the environment, the viable system filters specific characteristics it feels are necessary to satisfy its needs. By means of a self-organizing capability, the viable system re-qualifies them in order to create a context adequate to its own survival. This activity stems from an evaluation of relevance, i.e. the importance that the external viable system has in the perception of decision makers in terms of power of influence and control of resources critical for the survival and effective functioning of the governed system.

In sum, *vSA*, inspired by Beer's model, develops a systems thinking based mind-set characterised by several fundamental shifts of focus compared to the most common ones, which can be understood as follows: from parts and relations to interaction, from an internal focus on the system's operations to an external focus on the system's context, from a structural to a systemic perspective.

Thus, the main implications are:

- Dynamism (the development of structure in accordance with emerging changes);

- Contextualization (the search for viability through interaction with certain privileged entities), such as suprasystems that influence the system's survival);
- Constructivism (the decision maker's view of reality).

Therefore, the general scheme of the *structure-system* paradigm proposes a dual perspective of investigation (Barile and Saviano 2011a). In particular, it acts by focusing on:

- How the system/phenomenon is made up (defining the so-called *Structure-Based View*). A static and objective view, it is an approach which focuses on objects (such as parts, components, describing an analytical and reductionist approach) and on relations (relational view);
- How the system/phenomenon functions/behaves (defining the so-called *Systems-Based View*). A dynamic and subjective observation, it is a perspective that extends the view from the parts and relations (statically and descriptively observed) to the whole interaction (dynamically interpreted) process, drawing a systems view.

Central is the mentioned constructivist view, which highlights that different environments are extracted from the complexity of reality as well as different contexts from the same environment. Central is the notion of emergence (Morin 2007; Minati et al. 2006), which highlights that a system can emerge from a structure, and what is more important that different systems can emerge from the same structure and a system can emerge from different structures (Pels et al. 2014).

All these elements are summarised in five principles that define and shape the main concepts of *vSA*: Survival, Eidos, Isotropy, Ethos and Exhaustiveness, on which the building of a coherent body of general schemes is based:

- Survival: *A Viable system, living in a specific context, has the primary purpose of survival (survival, viability, life cycle).*
- Eidos: *The Viable system in an ontological sense may be observed from a dual perspective: the structure and the system (the binominal structure-system and static-dynamic).*
- Isotropy: *The viable system, in its behavioural sense, is characterized by two logically distinct areas: decision, linked to the governance function and action referred to the management function (the binominal think-act; idea-project-work).*
- Ethos: *The viable system, in its existential dynamics, is influenced in the pursuit of goals and in the achievement of objectives by the interaction with the suprasystems and subsystems from which and to which, respectively, elicits and provides guidelines and rules (openness, consonance, requisite variety).*
- Exhaustiveness: *In a viable system, all entities external to its structure are viable systems or components of an upper level viable system (the binominal internal-external, recursion).*

As one of the key implications that become apparent when looking at a system through the lens of *vSA*, the system's boundaries vanish at the precise moment the system starts its existence because the environment is being subjectively and constantly construed by the observer as a context.

Therefore, the system's survival depends on the decision maker's capability of interpreting the environment in terms of constraints (what we should do, *compliance*), possibilities (what we may do, *exploration*), and opportunities (what we can do, *exploitation*). This triad, like many others that easily emerge in the *vSA* view, should be embedded in any framework of management decision making because it is at the basis of survival of the system, which depends on the survival of the context in which it lives and aims to remain viable, and introduces a *sustainability* perspective. This aspect, we discuss in the next section, is fundamental in the *vSA* view of reality and highlights how, the discussion of complexity, and the survival of systemic entities in a complex scenario, implies inevitably, discussion about sustainability.

3.2.2 The vSA *View of Complexity*

Existing management models, techniques and tools are more or less codified solutions to experienced problems, whose underlying functioning rule has been understood. *vSA* defines such models 'schemes of synthesis' with reference to problem solving. However, with growing conditions of complexity, the capacity of managers is constantly required to face problems and situations never experienced previously. Decision making is therefore a context in which decision takes place without the interpretative support of experienced models, techniques and tools (Barile 2009a, b). Thus, while problem solving can be addressed with the aid of the patrimony of knowledge of existing theories, models, techniques and tools (i.e. schemes of synthesis), the complexity of decision making has also to be dealt with taking into account such patrimony.

As it will be highlighted in next sections, a fundamental aspect emerges from the *vSA* interpretation of complexity: a 'positive' view of the same as a condition thanks to which the decision maker is pushed to search and explore the unknown so feeding his/her knowledge potential and enriching the patrimony of general schemes developed over time, hence reinforcing his/her dynamic capabilities to face new problematic contexts.

3.2.2.1 The *vSA* '4Cs' Curve

Complexity is defined as a situation whereby decision makers do not possess relevant schema to enable them to understand and govern experienced events. Such situations common not only in managerial fields also characterize everyday life. When the level of complexity rises, our available schema requires adjustments to deal effectively with the same. To understand this focal point more clearly, it

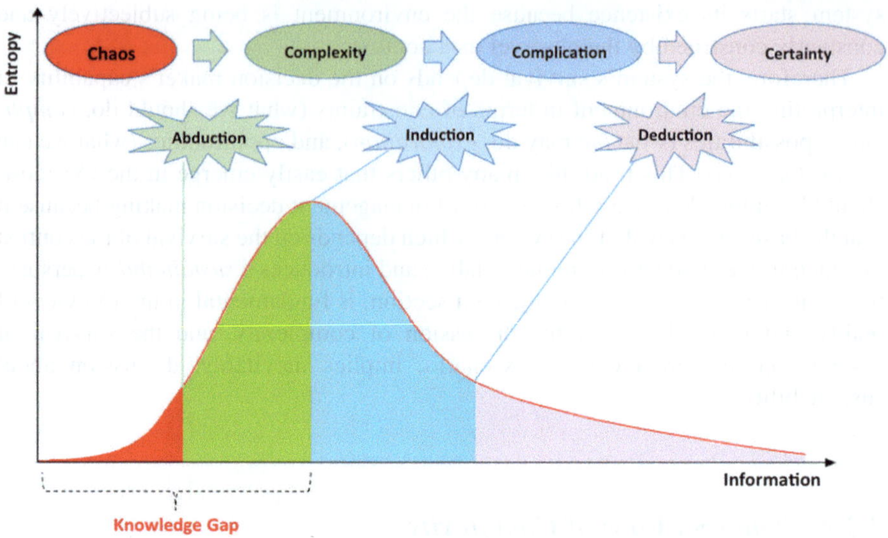

Fig. 3.2 The 4Cs curve. Source: Adapted from Barile (2009a, b, p. 53) and Barile et al. (2015a, b, p. 185). www.asvsa.org

might be useful to trace the four steps that define the process of decision making and knowledge acquisition and systemizing in a balanced way the entropy stemming from the binominal information-decision.[3] Dealing with an entirely new problem is a quite familiar experience for most individuals. Knowingly (or more often than not, unknowingly) they pass through four phases: from chaos, to complexity, to complication, to certainty. The steps detailed in the 4Cs curve are set out below in Fig. 3.2 (Barile 2009a, b).

The key of the curve pivots on recognizing that a problem appears simple or complex to a decision maker depending on the variety of knowledge in (his/her) possession, i.e. the information variety (Barile 2009a, b). Clearly, however, it is not merely a question of information. The knowledge required to effectively deal with a problematic situation does not only consist of units of information but also interpretation schemes and categorical values. Interpretation schemes, in particular, are fundamental to such 'understanding' i.e. implying the identification of the rule that governs and explains the observed phenomenon. Categorical values, instead, act by emotionally (given that it is a psychological process) directing interpretation and subsequently decision making towards resolution pathways consonant with an accepted values system (Barile 2009a, b; Barile et al. 2015a). Such interpretation, in which the decision maker's perspective is central, stands on a constructivist view of knowledge and decision making processes (von Glasersfeld 1984, 1989, 1995;

[3]The problem itself is neither chaotic nor complex or simple, rather, it is the decision maker with his/her personal information variety that identifies the correct schema to interpret, understand and solve the problem.

Vico 2008). This results in a rejection of approaches that aim to 'objectify' complexity by reducing it to mere calculation (Barile 2009a, b; Barile and Saviano 2011b). In this respect, a quantitative approach aimed at measuring complexity would inevitably fail in grasping the subjective dimension and would only provide a description of phenomena observed.

The law behind this curve, therefore, is easy to explain and relies on the concept of *information variety* (Barile 2009a, b). Whenever people have to face unfamiliar problematic situations, they do not possess relevant and consolidated interpretation schema. Therefore, they start to collect information and attempt to shape into or adapt the problem to their own set of acquired past schema. However, as the problem is quite unfamiliar, the lack of schema inhibits a linear interpretation and the creation of a unique and comprehensible corpus to interpret the binominal context-problem. New information does not help to clarify the problem but complicates the situation even more; the more information collected, the more entropy is experienced. The area in question shown in red in the chart is denominated *chaos*.[4]

In chaotic contexts, individuals have to trust their intuition and cognitive elaboration capabilities that we can call *abduction*, when finalised to the achievement of a specific aim. Abduction (the first noteworthy stage of the process) defines the shift from chaos to *complexity* (green area). At this stage, growing entropy and confusion are still in act but at a decreasing rate. However, this is not the problem solving stage nor that of creating adequate schema. It is what we call the 'eureka' stage lighting up in the distance.

The second stage takes place when after performing an *induction* phase, abductive eureka becomes an experimental hypothesis. This phase is that in which the subject may have found a schema—a specific mental order through which to organize the information that appears useful to explain and to solve the problem—and inverts the curve orientation toward consolidation of the potential solution pattern. The passage from complexity to *complication* is connoted by an inversion in the direction of the curve, which is now decreasing (blue area).

Intervention in the third stage (*deduction*) stresses the shift from complication to *certainty* (violet area). In this phase, new information is useful to better frame the solution found. In this sense, clarity derives from the definition of a problem and fitting to it interpretation schema.

An analysis of this model helps to better define the difference between problem solving and decision making. As Wolfram points out, "things seem complex if we don't have a simple way to describe them [...] there may be a simple underlying rule for something—even though the thing itself seems to us complex" (Wolfram 2008, pp. 131–134).

[4]It is worth noting that, to face a situation of chaos, creativity can be very helpful. Through creativity, individuals can blend values in order to generate new insights. See Barile, Saviano, Iandolo, 2012.

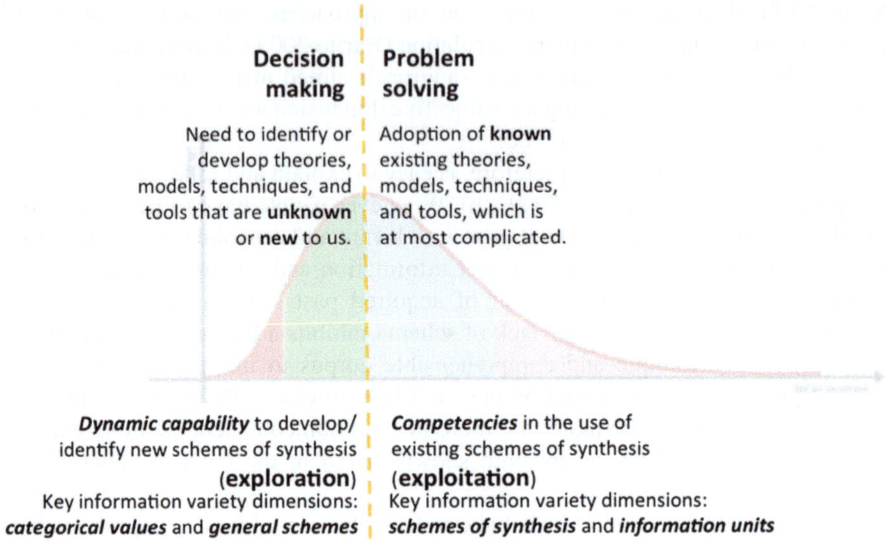

Fig. 3.3 Distinguishing decision making and problem solving. Source: Elaboration on Barile (2009a, b, p. 53). www.asvsa.org

3.2.2.2 Management Between Decision Making and Problem Solving

In the light of the above, decision making in condition of complexity implies the need to identify or develop theories, models, techniques, and tools that are currently unknown or new to us. It is the ability to identify and develop new interpretation schema and characterises the first two areas of the 4Cs curve (chaos and complexity), as shown in Fig. 3.3. Problem solving, on the other hand, is characterized by the adoption of (known) existing theories, models, techniques and tools and is represented by the second two areas of the 4Cs curve (complication and certainty). It is possible to define problem solving as the ability to use existing interpretation schema. The problem decision-makers face is not the availability of information but rather the knowledge gap experienced resulting from the lack of interpretation capabilities.

We should specify that in the managerial field, the set of organizational choices could refer to two different strands, depending on the context conditions that characterise the decision process. In this particular case, management decisions and governance decisions.

Management decisions are undertaken in contexts of complication or certainty and are characterised by available laws, rules, customs and interpretation schemes. In this sense, decisions are made generally based on models of experience involving the problem-solving realm (complication or certainty). Governance decisions, in contrast, are generally taken in conditions characterised by unavailability of shared laws, rules and customs and, subsequently, of reliable interpretation schemes (complexity if not chaos). Decisions are made based on feelings, values and desiderata (decision-making realm).

The pitfalls linked to decision making in the managerial field are quite evident. Management tends to believe that there is an optimal solution for every problem, ending up by adopting a problem solving approach in dealing with governance issues, which relate on the contrary, to the realm of decision-making, noting the growing inadequacy of models, techniques and tools already at their disposal.

A wider view of the notion of complexity is required to comprehend the steps to grasping a dynamic and subjective systems view of management. To clarify this point, we argue that complexity is not entrenched in the context or in the phenomenon itself, but is the (subjective) condition of the decision maker's cognitive process. An individual is influenced and affected by his/her interpretation schemes and by the emotions roused from perception of the context. Complexity manifests itself when interactions, emerging from the relations in a specific process, appear to the observer not to follow clear-cut behavioural rules.

In this sense, it would be interesting to analyse such interactions in depth to shift from an objective to a subjective view of complexity. According to Rullani, complexity is characterised by three dimensions: variety, variability and indeterminacy. In distinguishing between simple and complex, Rullani argues that we can call a problem, a solution, a context (in which the problem is to be found), a point of view (with which it is observed) complex if we recognize a high degree of variety, variability and indeterminacy for potential situations and events (Rullani 2008, p. 76).

Starting from the top, the concept of variety is the first level and can be defined as the potential set of modifications with which a phenomenon appears to the observer's eyes. At this stage, elements such as the characteristics and the connotations of the observed phenomenon and the descriptive classifying attributes are considered. As already stated, however, variety is a static (synchronic) representation of the set of elements that characterize the phenomenon under observation. Clearly, the fact that we can 'see' this set implies that it is more or less known to us; therefore, it can appear to us at most complicated, or at least not simple. What qualifies complexity requires the shift to a deeper level of understanding, through the adoption of a dynamic view of how phenomena change over time, i.e. its variability.

The most important thing to stress is that variability is a diachronic construct thus implying the concept of time: we can say that it gives movement to variety through change. This change implies the modification of the descriptive attributes of the phenomenon or the emersion of new ones linked to a sum of factors unfamiliar to the observer. In this respect, it is generally thought that the wider the variety of an observed phenomenon, the higher the probability that a new emerging variety will be experienced, concluding that variety and variability indicate emerging complexity. Although this is probably true in reality, such a conclusion is not a commonplace. If the interpretation scheme is right, i.e. if we understand the rule that governs the observed phenomenon, increasing variety and variability do not affect our comprehension and we follow the curve without varying entropy.

If, instead, incoming variety (information units) implies a change in the behaviour of phenomena that appears incomprehensible, it means that new rules are governing it and we need to rethink our interpretation schemes accordingly. This generates complexity and the curve starts to rise again.

In this scenario, indeterminacy occurs. In particular, it is possible to imagine indeterminacy as a bridge between the consciousness of complexity and the definition of a set of processes and schema in order to govern it. The shift from variability to indeterminacy is much deeper compared to the process outlined above. Through indeterminacy, in fact, we express the subjective condition of decision makers, the need they have to frame unknown phenomena, and the shift from a situation of complication to one of complexity.

These dynamics can never be generalized, as they show completely different trends depending on the actors (e.g. decision makers) involved in the process.

3.2.2.3 The *vSA* "T-Shaped" Knowledge Between Exploration and Exploitation

A useful declension in managerial terms of the *vSA* distinction between decision making and problem solving can be found in the "T-Shaped" framework (Hansen and Von Oetinger 2001), which offers a rethinking of the set of skills managers should possess in order to face current (and future) challenging scenarios. Ever more widely shared learning pathways define education programs and didactic structured approaches on the idea of "T-Shaped" professionals, combining grounded vertical expertise, in one or more disciplines or systems with boundary-crossing horizontal capabilities between disciplines and systems to deal with the various and emerging problematic contexts. The *vSA* version of the T-Shaped concept (Barile et al. 2012, 2014 Barile and Saviano 2013; Saviano et al. 2016) is configured on the information variety representation, made up of information units, general interpretation schemes, schemes of synthesis and categorical values.

In Fig. 3.4, an elaboration of the 'T-Shaped' idea, which integrates key elements of the 4Cs curve and the Fig. 3.3, shows how the dynamic capabilities and competencies are developed and act in the knowledge process that characterizes decision making and problem solving. So re-interpreted, the model underlines the relevance of the general schemes in grounding the dynamic capabilities (Teece et al. 1997) of the decision makers to face the emerging complexity of a changing scenario by adopting an exploration approach (March 1991).

In particular, the general schemes enable a boundary-crossing movement across the various problematic contexts experienced and acts as a bridging of knowledge between the various disciplinary domains, progressively enlarging the knowledge potential of the decision maker. The schemes of synthesis represent structured and contextualized knowledge that is developed when a solution is devised for a specific problem; they qualify the system's set of competencies and a knowledge approach

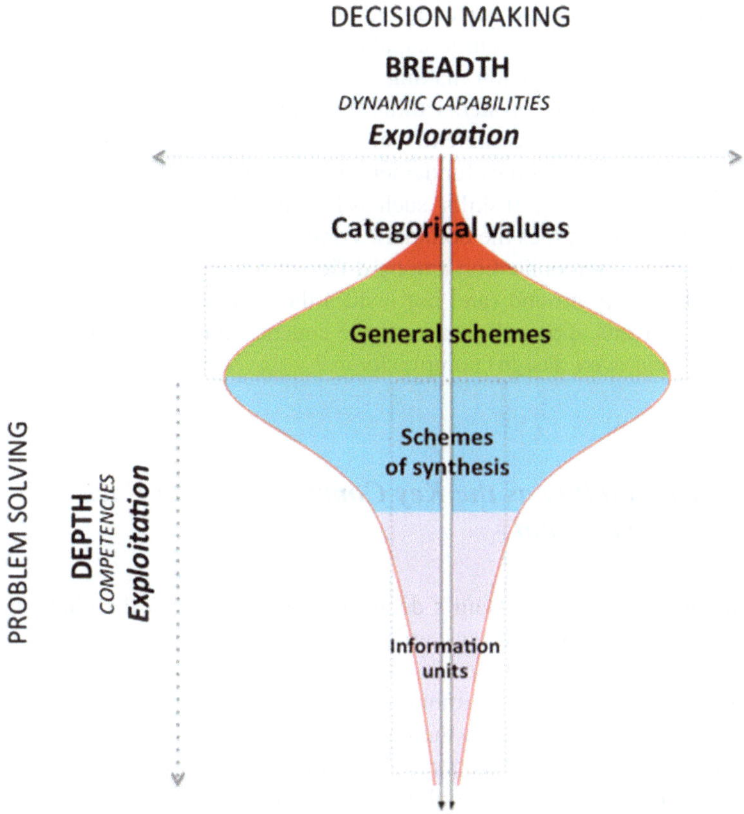

Fig. 3.4 The 'T-Shaped' knowledge view of the *vSA*. Source: Elaboration from Barile (2009a) and Saviano et al. (2016). www.asvsa.org

based on exploitation. The information units represent the data possessed by the system related to the problematic context experienced. Integrating the 4Cs model into the *vSA* T-shaped representation makes clear the role of the categorical values that represent sets of values, strong beliefs, convictions—which characterize the identity and personality of the decision maker—and act by directing his/her behavioural choice in condition of chaos or complexity, when information units are lacking or current schemes of synthesis are not useful to decision (Barile et al. 2014b).

The basic concept of "T-Shaped" is that while disciplinary, professional and technical knowledge is essentially procedural and based on adopting well-known models, techniques and tools (schemes of synthesis), the knowledge that develops and is developed thanks to the dynamic capabilities rather than technical is methodological in nature in the most common meaning of method as the 'way of doing' (Barile et al. 2017).

Management methods are currently lacking not much technical or instrumental competencies, but mainly dynamic capabilities of dealing with the emerging and growing complexity of decision making. At this level, the contribution of systems thinking is incomparable: systems thinking supports the enlarging of the view necessary to deal with the complexity of the current scenario, and the *vSA* offers the basic set of general schemes fundamental to explore the "unknown" by leveraging the decision makers' 'soft skills' such as lateral and critical thinking. Moreover, systems thinking overcomes the boundaries drawn by the divides of disciplines and returns an idea of interconnection among different fields of knowledge. The possibility to build reality around (and not inside) the boundaries of disciplines is the main strength of this approach: it adopts the constructivist perspective not to face but to frame (and even design) complexity.

3.2.3 Sustainability as the Key Complex Issue to Face in Management

The enlarged view of the system's decision maker opens up the challenges that must be addressed to ensure the system's viability over time. The notion of viability is central in the framework of the *vSA* and its relationship with that of sustainability has gained increasing attention among *vSA* scholars interested in investigating the survival condition of systems. In the next sections, the notions of viable systems' survival and viability, which are strongly related to that of complexity, are under focus highlighting their intrinsic link to the concept of sustainability.

3.2.3.1 From System Survival to System Sustainability

As a fundamental aspect highlighted by systems thinking and *vSA* in the shift of focus from the objective analysis of the system's structure to the system's context, the notion of survival in the *vSA* implies understanding the conditions through which a viable system interacts and frames its environment, by defining the most appropriate relational strategy in order to harmonize needs and expectations of relevant suprasystems and subsystems.

According to the *vSA*, in fact, in order to survive, a system needs to establish relationships with other viable entities (suprasystems) that hold the resources necessary for its survival and effective functioning. Hence, to gain access to needed resources, the system's decision maker must involve suprasystems (and subsystems) within a network relationship whose effectiveness depends on the capability of harmonizing multiple (often diverging) interests, engaging actors in a process of value *co-creation*.

In order to define the most appropriate relational strategy, the system's decision maker evaluates the suprasystems' degree of relevance for the system's functioning and viability. Relevance is evaluated in terms of how critical the suprasystems' resources are (*criticality*) and how capable of influencing the system's dynamics they are (*influence*).

In order to harmonize the needs and expectations of suprasystems and subsystems, the system's decision maker must govern a delicate equilibrium between two forces under which decisions are made:

- *Impulsive force* (competitiveness) that defines efforts towards the continuous improvement of the system's own performances 'against' the context;
- *Field force* (consonance) that defines indulgence 'towards' the context, aspiring to harmonious relationships with sub and suprasystems to achieve collective potential, an ideal sense of belonging to the system.

Therefore, consonance refers to the relational compatibility/complementarity between two or more actors that interact for the purpose of an emerging system. It expresses the potential for value to emerge.

The result of a virtuous interaction (harmonization) between the two drivers is the resonance. When consonant actors interact, they generate resonance (value co-creation): the process and the outcome that emerge from the developed synergies. Resonance makes possible the existence of a collective consciousness, which (in turn) makes desired future scenarios come true (Barile 2009a, b).

vSA, therefore, drives us beyond the basic issues of management. In this sense, returning to the first areas of enquiry into systems thinking ("*the structure and operations of living systems and their relationship with the environment*"), we can shed new light on the way viable systems (individual and organizations) survive in the socio-economic and natural environment.

The focus on context, which characterizes *vSA*, becomes the key for a new understanding of the relationship between (viable) systems and their environment. The *vSA* notion of viability, in particular, intended not so much as autonomous existence in the environment, but as harmonious existence, is the key.

In the *vSA* perspective, in fact, viability is the result of strategic choices on the part of decision-makers who, by virtue of their skills of reading the context and the proper interpretation of the set of laws, rules and regulations that constrain and limit the range of possible actions, define pathways of *sustainable development* of the system. In this view, the concept of sustainability, through that of survival, results intrinsic in the concept of viability. Adopting a sustainability perspective, therefore, helps to specify the conditions of survival for the system, also in terms of probability and risks.

In essence, the *vSA* helps to see the strong connections between the survival of individual systems and the survival of their environment: *to increase the probability of its own survival, each system should contribute to safeguard the conditions for survival of the environment in which it lives.*

Although this may appear an obvious consideration, routine evidence of the way individuals and organizations live shows how far we are from a condition of synergistic relationship between individuals and organizations and their environment. It is no coincidence that the notions of sustainability and sustainable development have been debated at length at almost every level (scientific, social, political, cultural, etc.) of human life without evidencing a true transition towards a more sustainable world. This suggests that, despite massive efforts and a great variety of initiatives, results are still unsatisfactory. Hence, although relevant progress has been made, especially in the technological field, sustainability is still a complex issue for decision makers, hard to put in practice (Golinelli and Volpe 2012). Probably, what is required and still missing, is a true paradigm change toward sustainability, which, given the apparent systemic nature of the dynamics that impact on sustainability, cannot be achieved without adopting a systems approach. Hence, a key for deciphering the complexity of sustainability is required, and system thinking can offer a relevant contribution (Barile et al. 2014b, 2016b; Golinelli et al. 2015).

A recursive systemic link, in fact, between the various levels of survival—from individual systems to the planet as a whole—clearly appears, highlighting the necessity for decision makers to shift from an *ego*-systems to an *eco*-systems view based on the principle that any living (and viable) system should not live at the expenses of other systems. Therefore, the link between the perspectives of viability and sustainability, which appears intrinsic in the *vSA* paradigm, suggests an in-depth understanding of sustainability through the adoption of a systems perspective, and its implications. This next section discusses in depth the concept of sustainability starting from the most common reference to its triple environmental, social and economic dimensions, on which the well-known *Triple Bottom Line* framework is based (Elkington 1997).

3.2.3.2 Deciphering the Complexity of Sustainability Through Systems Thinking

A viable system arises from the will of a decision maker to organize resources in order to achieve a specific goal. This system is viable thanks to the capability of self-determination, by redesigning its own structure and defining new strategies with the specific aim of survival in the context in which it acts. Resulting from such self-determination, the system has responsibility for the various categories of actors with which it has defined relations, throughout its evolutionary dynamics. Therefore, through the transformation of choices made, decision makers act within a specific range of possibilities, on a far wider environmental scale. Hence the necessity to develop a holistic view of the dynamics generated by interactions between the three dimensions (environmental, social and economic), which, according to the Triple Bottom Line framework, define sustainability.

The model developed on the triad *people, profit and planet*, synthetizes the sustainability perspective in three main dimensions: society, environment and economy.[5] The adoption of a sustainability perspective is an intrinsic necessity

[5]The schema behind this triad is that of Aristotle: *essence, potency and act*. It also recalls the modal logic triad of *necessity, possibility and effectivity*.

for the identification of conditions for survival of the system in its viable context. However, looking at the current situation, we see evidence of an imbalance where the economy dominates the other two dimensions. In this sense, the promotion of a change, a transition to a situation in which environment and society will recover the lost equilibrium in their interaction would be required.

If we look at the reality surrounding us as a set of areas, each of which anthropomorphically defined and, therefore, ever changing, the definition and the representation of sustainability as the outcome of interactions between its three dimensions suffers from the limits of an approach that is considered reductionist. Sustainability, in fact, identified as the static determinant of the intersection of three areas, seems to be the consequence of a mechanism that could be defined deterministic (cause-effect): the triple bottom line model which affirms that if interaction between the three areas occurs, *sustainable intersection will emerge*, which could be assured, and, therefore, in theory, always actionable.

The interpretation proposed however, tries to overcome the limits of the simple 'sum' of the identified parts, focusing on the virtuous dynamic of interaction that these dimensions can generate. The shift implies the overcoming of a reductionist view, which tends to focus on single areas, in favour of a view that takes into account the overall processes involving them, moving towards their integration based on the understanding of how they interact (Barile and Saviano 2011b; Barile 2013).

The integration of the three different domains and perspectives (environmental, social and economic) stated by the Triple bottom line representation of sustainability, in actual fact refers to two typical *Complex Adaptive Systems* (CAS) (Folke et al. 2002; Rammel et al. 2007):

- *Social-Ecological Systems* (SESs) (Ostrom 2009) focused on humans-nature coupled systems, in which, however, the environmental perspective dominates;
- *Socio-Technical Systems* (STSs) (Trist 1981) focused on humans-technology coupled systems, in which, however, the economic (and engineering) perspective dominates.

The complexity of the interconnected dynamics of socio-technical-ecological systems implies many trade-offs and emerging outcomes to deal with through boundary crossing interaction. In this sense, the challenge is how to integrate these different domains (hence, related perspectives), in order to recover a unitary view. As suggested by Fig. 3.5, it is possible to see that in both the CAS, there is a social component. Moreover, they are both 'systems'. Thus, the social component can act as a bridge between the economic and the environmental components, which differentiate the two CAS, and systems thinking can support this process. The T-shaped notion introduced above suggests that it is necessary to create an interdisciplinary context between social sciences, environmental sciences and economic sciences (Saviano et al. 2016): the social sciences can act as bridging element and the systems approach provide the horizontal general schemes required

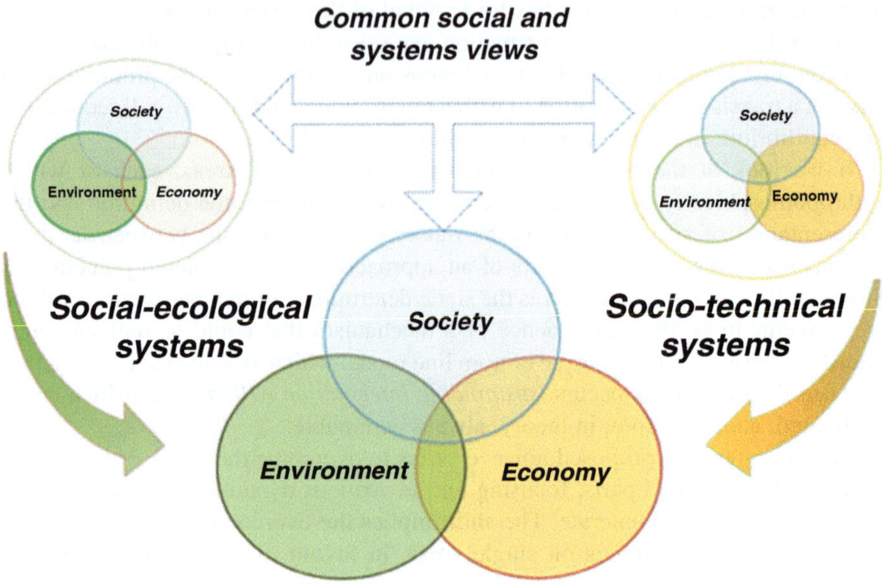

Fig. 3.5 Bridging socio-ecological systems and socio-technical systems through the social and systems views. Source: Adapted from Golinelli et al. (2015). www.asvsa.org

to cross the boundaries between disciplines, using a common set of knowledge and a common code of language that facilitates understanding.

This view implies that, in order to implement a process that leads to the recovery of conditions for sustainability, we need to imagine that the three areas can contribute substantially to the development of a single evolutionary process, and that the criterion for the verification of sustainability is applied only to this process.

By adopting a 'complexity' perspective, as proposed in Fig. 3.6, the 4Cs Curve helps to understand the 'law' that should govern interaction between the three domains of sustainability. Essentially, the 'environment' area should be conceived as a set of 'resources', but also of constraints, on which to base the emergence of the above process (*necessities*). Without understanding the laws of systemic functioning, this complex of resources would appear chaotic. It is in the 'social' area that the conception of 'potential ways', i.e. applicable rules in order to take advantage of the resources of the first area, should be performed thanks above all to science. In this sense, this area becomes a potential set of all the alternative use of what is available (*possibilities*). Lastly, the 'economy' area is the final element of this process: based on the resources of the first area and on the rules of the second area, it represents the area of effectiveness and includes the set of real and objective states of implementation of what is possible and made available by the previous areas (*actuations*).

From a deeper perspective, therefore, the analysis in terms of sustainability of the realization processes of the economic area is affected, more implicitly than explicitly, by constraints and rules imposed by the previous areas. In other terms, it

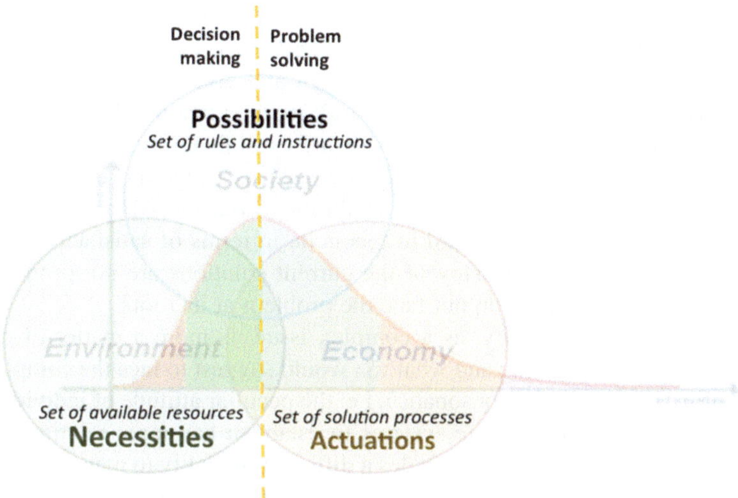

Fig. 3.6 A 'complexity' view of sustainability. Source: Authors' elaboration. www.asvsa.org

is not sustainable what is economically feasible, but only what may be 'justified' by laws, rules and social customs, as well as what does not exploit the availability of resources.

As suggested by the representation, on the one side, the variety to manage—naturally available and/or socially created—increases; on the other side, is reduced: it clearly appear the complexity of decision making in the former context and the 'simple' complication of problem solving in the latter context.

The conditions of harmonic interaction between the three domains of sustainability are the key. Therefore, the key to overall equilibrium is in the coherent functioning of the whole. Any kind of separation between what is unitary and interconnected would produce distortions and disequilibrium.

3.2.4 Concluding Remarks: The Interpretative Power of Systems Thinking

What emerges from the proposed discussion about complexity and sustainability in management is the revealing power of the general schemes of systems thinking that the *vSA* has re-explored with the aim of providing a possible 'method' organized a coherent framework to support management decision making in condition of complexity.

The discussion about the interpretation of the sustainability framework provides an example of the complexity of the challenges in play. A reductionist view of the issues that must be faced would not help to accomplish a real transformative change necessary to progress towards sustainability. So far, the major initiatives adopted by governments, society and economy reveal that a systems view is still lacking and actions at global and local level do not seem to effectively converge. The variety of problematic contexts and issues to face is high and gives rise to many trade-offs hard to solve. In the business context, several decision making dilemmas disorient companies wanting to promote the shift towards sustainability: it would be necessary to harmonize targets hard to reconcile such as economic efficiency, social effectiveness, and environmental sustainability (Saviano 2016). Many of these trade-offs imply 'possibilities' hard to reconcile in terms of satisfactory solutions for each actor involved and many of the current solutions are not producing the desired effects because they do not face the problem at its roots.

In *vSA* terms, sustainability, as a complex issue, is more a matter of decision making than problem solving and what *vSA* would suggest to face the complexity of sustainability is a tension to consonance, i.e. the peculiar attitude of individuals and organizations designed as *viable* systems to establish harmonic relationship with the context. In other words, *vSA* provides a different modality to manage the flow of different will and choices that define decisions under uncertainty: designing complexity to balance environmental, social and economic perspectives means to be aware of the need to change premises and questioning beliefs, to transform a set of resources and rules (necessity and possibility) in a consonant set of solutions (actuations).

Therefore, each viable system has to shape its variety to comprehend the environment, has to be adaptive to survive, has to be sustainable to co-create value, develop and evolve.

References

Ashby, R. W. (1968). Principles of self-organizing system. In W. Buckley (Ed.), *Systems research for behavioral science* (pp. 116–117). New York: Aldine Transaction.

Barile, S. (Ed.). (2000). *Contributi sul pensiero sistemico in economia d'impresa*. Salerno: Arnia.

Barile, S. (Ed.). (2005). *L'impresa come sistema. Contributo sull'Approccio Sistemico Vitale (ASV)*. Torino: Giappichelli.

Barile, S. (2009a). *Management Sistemico Vitale*. Torino: Giappichelli.

Barile, S. (2009b). The dynamic of information varieties in the processes of decision making. In *Proceedings of the 13th World Multi – Conference on Systemics, Cybernetics and Informatics, WMSCI*, Florida.

Barile, S. (Ed.). (2013). *Contributions to theoretical and practical advances in management. A viable systems approach (vSA)*. Roma: Aracne.

Barile, S., & Saviano, M. (2011a). Foundations of systems thinking: The structure-system paradigm. In Various authors, *Contributions to theoretical and practical advances in management. A viable systems approach (vSA)* (pp. 1–25). Avellino: International Printing. www.asvsa.org

Barile, S., & Saviano, M. (2011b). Qualifying the concept of systems complexity. In Various authors, *Contributions to theoretical and practical advances in management. A Viable Systems Approach (vSA)* (pp. 27–63). Avellino: International Printing. www.asvsa.org

Barile, S., & Saviano, M. (2013). Dynamic capabilities and T-shaped knowledge. A Viable Systems Approach. In S. Barile (Ed.), *Contributions to theoretical and practical advances in management. A viable systems approach (VSA)* (Vol. II, pp. 39–59). Roma: Aracne.

Barile, S., Pels, J., Polese, F., & Saviano, M. (2012). An introduction to the viable systems approach and its contribution to marketing. *Journal of Business Market Management, 5*(2), 54–78.

Barile, S., Carrubbo, L., Iandolo, F., & Caputo, F. (2013). From 'EGO' to 'ECO' in B2B relationships. *Journal of Business Market Management, 6*(4), 228–253.

Barile, S., Saviano, M., & Simone, C. (2014). Service economy, knowledge and the need for T-shaped Innovators. *World Wide Web, 18*(4), 1177–1197.

Barile, S., Sancetta, G., & Saviano, M. (2015a). *Management. Il modello sistemico e le decisioni manageriali* (Vol. 1). Torino: Giappichelli.

Barile, S., Saviano, M., Polese, F., & Caputo, F. (2015b). T-shaped people for addressing the global challenge of sustainability. In E. Gummesson, C. Mele, & F. Polese (Eds.), *Service dominant logic, network and systems theory and service science: Integrating three perspectives for a new service agenda*. Napoli: Giannini.

Barile, S., Lusch, R., Reynoso, J., Saviano, M., & Spohrer, J. (2016a). Systems, networks, and eco-systems in service research. *Journal of Service Management 27*(4), 652–674.

Barile, S., Saviano, M., Iandolo, F., & Caputo, F. (2016b). La dinamica della sostenibilità tra vortici e correnti: un modello a Triple Elica. In Various authors, *Atti del Convegno Nazionale dell'Accademia Italiana di Economia Aziendale 2015* (ed). Bologna: Il Mulino.

Barile, S., Saviano, M., Calabrese, M., & La Sala, A. (2017). In search of a possible evolutionary principle of management theory and practice. In *Science with and for Society – Contributions from Cybernetics and Systems. WOSC 2017 Book of Proceedings* (forthcoming).

Beer, S. (1972). *Brain of the firm*. London: The Penguin Press.

Beer, S. (1974). *Decision and control*. London: John Wiley & Sons.

Beer, S. (1975). *Fanfare for effective freedom, in platform for change*. Chichester: Wiley.

Beer, S. (1985). *Diagnosing the system for organisations*. Chichester: Wiley.

Bogdanov, A. (1912). *Tektology: The universal organizational science* (Vol. I). St. Petersburg.

Buckley, W. (1967). *Sociology and modern systems theory*. Englewood Cliffs, NJ: Prentice Hall.

Buckley, W. (1968). *Systems research for behavioral science*. Chicago, IL: Aldine Publishing Company.

Buckley, W. (2008). Society as a complex adaptive system. *Emergence: Complexity and Organization, 10*(3), 86–112.

Ceruti, M. (1986). *Il vincolo e la possibilità*. Milano: Feltrinelli.

Clark, M. C. (1993). Transformational learning. In S. B. Merriam (Ed.), *An update on adult learning theory, new directions for adult and continuing education* (Vol. 57, pp. 47–56).

Cyert, R. M., & March, J. G. (1963). *A behavioral theory of the firm*. Oxford: Blackwell Publishing.

Elkington, J. (1997). *Cannibals with forks. The triple bottom line of 21st century*. London: New Society Publishers.

Emery, F. E. (Ed.). (1969). *Systems thinking: Selected readings*. Harmondsworth: Penguin.

Espejo, R. (1989). A method to study organizations. In R. Espejo & R. Harnden (Eds.), *The viable system model: Interpretations and applications of Stafford Beer's VSM*. Chichester: Wiley.

Espejo, R. (1993). Management of complexity in problem solving. In R. Espejo & M. Schwaninger (Eds.), *Organizational fitness. Corporate effectiveness through management cybernetics*. Frankfurt: Campus Verlag.

Folke, C., Colding, J., & Berkes, F. (2002). Synthesis: Building resilience for and adaptive capacity in social-ecological systems. In F. Berkes, J. Colding, & C. Folke (Eds.), *Navigating social-ecological systems: Building resilience for complexity and change* (pp. 352–383). Cambridge, UK: Cambridge University Press.

Golinelli, G. M. (2000, 2005). *L'approccio sistemico al governo dell'impresa* (Vol. I, I and II ed.). Padova: CEDAM.

Golinelli, G. M. (2005). *L'approccio Sistemico Vitale (ASV) al governo dell'impresa*. CEDAM.

Golinelli, G. M. (2010). *Viable systems approach (VSA). Governing business dynamics*. Padova: CEDAM/Kluwer.

Golinelli, G. M., & Volpe, L. (2012). *Consonanza valore, sostenibilità. Verso l'impresa sostenibile*. Padova: Cedam.
Golinelli, G. M., Barile, S., Saviano, M., Farioli, F., & Yarime, M. (2015). Towards a common framework for knowledge co-creation: Opportunities of collaboration between service science and sustainability science. In E. Gummesson, C. Mele, & F. Polese (Eds.), *Service dominant logic, network and systems theory and service science: Integrating three perspectives for a new service agenda*. Napoli: Giannini.
Hannan, M. T., & Freeman. (1977). The population ecology of organizations. *American Journal of Sociology, 82*, 929–964.
Hansen, M. T., & Von Oetinger, B. (2001). Introducing T-shaped managers. Knowledge management's next generation. *Harvard Business Review, 79*(3), 106–116.
Hinterhuber, H. H., Friedrich, S. A., Handlbauer, G., & Stuhec, U. (1996). The company as a cognitive system of core competences and strategic business units. *Strategic Change, 5*, 223–238.
Kahneman, D., & Tversky, A. (1982). The psychology of preferences. *Scientific American, 246*(1), 160–173.
Kast, F. E., & Rosenzweig, J. E. (1972). General systems theory: Applications for organization and management. *Academy of Management Journal, 15*(4), 447–465.
Katz, D., & Kahn, R. L. (1966). *The social psychology of organizations*. New York: Wiley.
Kuhn, T. S. (1962). *The structure of scientific revolutions*. Chicago, IL: University of Chicago Press.
Lorsch, J. W., & Lawrence, P. R. (Eds.). (1970). *Studies in organization design*. Homewood, IL: Richard D. Irwin.
Luhmann, N. (1990a). *Die Wissenschaft der Gesellschaft*. Frankfurt: Suhrkamp.
Luhmann, N. (1990b). The cognitive program of constructivism and a reality that remains unknown. In W. Krohn, G. Küppers, & H. Nowotny (Eds.), *Selforganization. Portrait of a scientific revolution* (pp. 64–85). Dordrecht: Reidel.
March, J. G. (1991). Exploration and exploitation in organizational learning. *Organization Science, 2*(1), 71–87.
March, J. G., & Simon, H. A. (1958). *Organizations*. New York: Wiley.
Maturana, H. R. (1970). *Biology of cognition* (Biological Computer Laboratory Research Report BCL 9.0). Urbana, IL: University of Illinois.
Maturana, H. R. (1975). The organization of the living: A theory of the living organization. *International Journal of Man-Machine Studies, 7*(3), 313–332.
Maturana, H., & Varela, F. (1980). Autopoiesis and cognition. *Boston Studies in Philosophical Science, 42*.
Minati, G., Pessa, E., & Abhram, M. (2006). *Systemics of emergence: Research and development*. New York: Springer.
Morin, E. (2007). Le vie della complessità. In G. Bocchi & M. Ceruti (Eds.), *La sfida della complessità*. Milano: Bruno Paravia Editori.
Ostrom, E. (2009). A General framework for analyzing sustainability of social–ecological systems. *Science, 325*, 419–422.
Parsons, T. (1971). *The system of modern societies*. Prentice-Hall, NJ: Englewood Cliffs.
Pels, J., Barile, S., Saviano, M., Polese, F., & Carrubbo, L. (2014). The contribution of vSa and SDL perspectives to strategic thinking in emerging economies. Managing service quality. *An International Journal, 24*(6), 565–591.
Philips, D. C. (1971). Systems theory – A discredited philosophy. In P. P. Schoderbek (Ed.), *Management systems*. New York: John Wiley & Sons.
Rammel, C., Stagl, S., & Wilfing, H. (2007). Managing complex adaptive systems—A co-evolutionary perspective on natural resource management. *Ecological Economics, 63*(1), 9–21.

Rapoport, A., & Horvath, W. J. (1968). Thoughts on organization theory. In W. Buckley (Ed.), *Modern systems research for the behavioural scientist*. Chicago, IL: Aldine Publishing Company.

Rullani, E. (2008). L'economia della conoscenza nel capitalismo delle reti. Sinergie, Italian. *Journal of Management, 76*, 67–90.

Saviano, M. (2016). Il valore culturale del patrimonio naturale nella promozione dello sviluppo sostenibile. Sinergie. *Italian Journal of Management, 34*(99), 167–194.

Saviano, M., Barile, S., Spohrer, J., & Caputo, F. (2016). A service research contribution to the global challenge of sustainability. *Journal of Service Theory and Practice* (forthcoming).

Simon, H. A. (1967). Motivational and emotional controls of cognition. *Psychological Review, 74* (1), 29–39.

Simon, H. A. (1991). Bounded rationality and organizational learning. *Organization Science, 2*(1), 124–134.

Simon, P., & De Laplace, M. (1995). *A philosophical essay on probabilities*. North Chelmsford, MA: Courier Corporation.

Taylor, F. W. (1914). Scientific management. *The Sociological Review, 7*, 266–269.

Teece, D. J., Pisano, G., & Shuen, A. (1997). Dynamic capabilities and strategic management. *Strategic Management Journal, 18*(7), 509–533.

Trist, E. (1981). The evolution of socio-technical systems. *Occasional Paper, 2*, 1981.

Various Authors. (2011). *Contributions to theoretical and practical advances in management. A viable systems approach (vSA)*. Avellino: International Printing.

Vico, G. B. (2008). La scienza nuova. In P. Rossi (a cura di). Milano: Biblioteca Universale Rizzoli.

von Bertallanfy, L. (1967). General theory of systems: Application to psychology. *Social Science Information, 6*, 125–136.

von Bertallanfy, L. (1968). *General system theory. Foundations, development, applications*. New York: Penguin University Books.

von Glasersfeld, E. (1984). An introduction to radical constructivism. In P. Watzlawick (Ed.), *The invented reality* (pp. 17–40). New York: Norton.

von Glasersfeld, E. (1989). *Linguaggio e Comunicazione nel costruttivismo radicale*. Milano: Metope Clup.

von Glasersfeld, E. (1995). *Radical constructivism: A way of knowing and learning. Studies in mathematics education series: 6*. Bristol: Falmer Press/Taylor & Francis Inc.

Wolfram, S. (2008). Complexity: 5 Questions. 12 November 2008, Automatic Press/VIP, ch. 24, Carlos Gershenson (pp. 131–134). http://www.stephenwolfram.com/media/five-questions-about-complexity/, download October 2016.

Websites

www.asvsa.org

Chapter 4
Innovation Policies: Strategy of Growth in a Complex Perspective

Bruna Bruno, Marisa Faggini, and Anna Parziale

Abstract The aim of this chapter is to highlight new understandings of innovation as an interactive process in relation to economic growth. The resulting ideas could be of considerable interest to innovation policy makers. Two impacts are of considerable potential importance. The first relates to the absorption of complex and evolutionary systems dynamics ideas into the study of innovation, and growth. The second relates to the synthesis of complex systems ideas with evolutionary models of innovation, and growth. Considering innovation as a complex multi-level process means that it is not possible to devise the context into independent ways and that it is not enough to provide policymakers with simple solutions, but it should help them formulate and address questions that are appropriate to the evolutionary and complex context within which they operate.

Keywords Innovation • Complex systems • Policy making

4.1 Introduction

In the past, many historians and economists[1] emphasised the crucial role of technical institutional change in the theory of economic growth. Schumpeter regarded the process of innovation as central to understand economic growth, highlighting that the innovator rather than the investor or the inventor represented the most sensitive individual figure in the economy.

In the globalisation era, this issue is more important for economic development. At present we are looking for rapid and radical technological change that underlies some of the key economic developments. These include: turbulence and instability in industrial markets, major changes in world shares of output, persistent growth

[1]Adam Smith did recognize the great importance of science and technology, even if List (1841) had strongly criticized him and other classical economists for what he perceived as their neglect of technology and skills, that's they did not consistently give it the prominence which List thought that it merited (Freeman 2002).

B. Bruno • M. Faggini (✉) • A. Parziale
University of Salerno, Fisciano, Italy
e-mail: mfaggini@unisa.it

S. Barile et al. (eds.), *Social Dynamics in a Systems Perspective*, New Economic Windows, DOI 10.1007/978-3-319-61967-5_4

rate differences between economy, dramatic changes in company structures and production organisation, persistent high levels of unemployment, and increasing internationalization and economic integration.

The speeding up of the rate of innovation implying changes in technology, international trade and political deregulation registered an intensification of competition both in sectors already involved in international trade and in formerly protected sectors becoming less sheltered than before.

New agents and organisations confront themselves with new problems; new ways are evolving for relating to others and to important issues; the communities adopt tools and techniques ranging from collaborative visioning and decision-making processes to innovative technologies.

In this context the ability of a country to sustain rapid economic growth in the long run is highly dependent on the ways in which it can deal with evolving environment, and on the effectiveness by which its institutions and policies support the technological transformation and innovativeness. Policy makers have a much more complex task. They have to co-ordinate and calibrate three different policy areas: policies affecting the pressure for change (competition policy, trade policy and the stance of general economic policy); policies affecting the capability to impose and absorb change (innovation policy and human resource development); policies aiming at caring about losers in the game of change (social policies and transfer of income to weak regions).

To perform these tasks, complex system supporting institutions and good economic policies is needed, to foster policy co-ordination among different sectoral policies and across different territorial governments. This requirement has to be met because innovation policy affects the capability to innovate, and it needs to be co-ordinated with policies affecting the transformation pressure and with policies affecting income distribution.

Science and technology policy can no further be thought only in terms of research policy. The current territorial division of policy responsibility is not sustainable in globalisation context. For example, European authorities are in charge of policies imposing a pressure for change, while national and regional authorities are left with the responsibility to promote and cope with change. The policy maker can no longer look to research programmes which aim simply at the development of new scientific and technological principles, and results, but it is increasingly necessary to focus also on the application and use of science and technology by companies and by society as a whole because we are living in globalisation era. In fact because of technology is a vehicle for the diffusion of information and knowledge across borders, technological developments have themselves been stimulated by the globalization of markets. Much more evident is the increasing emphasis on the social and economic relevance and impact of research, and on the factors which shape its impact; in particular in areas as work and employment.

Central questions of policy concern how do firms and governments develop strategies for science, technology and innovation; which are the relationships between public and private in the creation of technology; what is the proper role of the public sector in this area; what are the implications of increasing internationalisation for national science policies. Economists and engineers, and in particular politicians are devoting increasing attention to understanding why, how and where technological innovations are generated and how and in which way these can affect the growth of a country.

The traditional literature to the innovation looked at the innovation following a linear approach. The development, production and marketing of new technology are assumed to follow a well-defined time sequence which corresponds less and less to the realities of the innovation process (Fisher 2002).

However, over the last twenty years economic theories of innovation have changed deeply, abandoning the traditional linear view of innovation for adopting systemic, or as we said, a complexity and evolutionary approach.[2] This approach emphasizes the central role of feedback between phases of innovation and numerous interactions between science, technology and economy. All these assumptions provide a foundation for systems approach for the analysis of innovation processes. According to this approach, innovations are seen as part of larger process of knowledge production of economic relevance. The approach stresses that firms do not innovate in isolation, but in interaction with other organizations: other firms, R&D institutions, universities and other forms of producer services (Fisher 2002).

Therefore, the innovation is a process that involves, in each moment, many and different actors, their relationships and the economic context in which they act. Under these perspectives (Kline and Rosenberg 1986) innovation should be the result of dynamic interactions among heterogeneous elements. It should be represented by a model in which the various aspects of economic activity are linked together by multiple relationships of causality and feedback and they are all interdependent elements of the process of innovation.

The aim of this chapter is to highlight new understandings of innovation as an interactive process in relation to economic growth. The resulting ideas could be of considerable interest to innovation policy makers. Two impacts are of considerable potential importance. The first relates to the absorption of complex and evolutionary systems dynamics ideas into the study of innovation, and growth. The second relates to the synthesis of complex systems ideas with evolutionary models of innovation, and growth. Considering innovation as a complex multi-level process (Frenken 2016) means that it is not possible to devise the context into independent ways and that it is not enough to provide policymakers with simple solutions, but it should help them formulate and address questions that are appropriate to the evolutionary and complex context within which they operate.

Following these suggestions, the paper analyses the most recent studies framing the innovation in economic growth, according to an evolutionary approach, by highlighting the differences with neoclassical approach. In this respect we offer an innovative approach to the analysis of innovation policies focusing the attention on National Innovation Systems Theory and its further developments. Finally, section six concludes the paper.

[2]This issue was explored in detail in Lane and Maxfield (1997, 2005).

4.2 Innovation Policy

An important topic in the international debate concerns the necessity of an innovation policy. This request derives from important issues outlining the role of technological innovation in modern capitalism development and determining the competitiveness and growth of a country.

In this scenario and in particular in the globalization era, the link among technology, innovation and competitiveness cannot be understood by using the traditional approach that focuses on the individual firms. The diversity study is the key to understand that linkage; the question is no longer how to characterize the similarities between firms, but rather what sense to make of the difference among them (Metcalfe 1994).

The implementation of innovation policy is not simple, despite the widespread attention and efforts dedicated to innovation issues from researchers and policymakers. It not possible to theorize simple policy prescriptions and devise context-independent ways to sustain it but it is important to conceptualize innovation as a complex, multi-level process. The inventions performed and the theoretical framework are not always perfectly synchronized because they continually evolve; the relationship among these actions involves different processes and institutional levels and they occur on different time. So, it could happen that the actors that are responsible to guide and implement policy are generally different from those that program concrete policy measures.

Furthermore, some goals of innovation policy would need the implementation of coordinated actions involving multiple policy fields and because policy programmes, once established, develop and continue over time, there is the risk that interventions may overlap by hampering the effectiveness of policies (Lane and Maxfield 2005).

From this, it follows the reject of the assumptions of neoclassical economic approach—representative agent, rationality axioms, stable equilibrium—and the understanding of behaviours in evolutionary context outlining the importance of history and the significance of small difference in behaviour. Of course, in this context also the innovation policy has to change perspective. The policy decisions are usually concerned with matter of details (single country, single firm, single agent).

Up to now, in fact, little importance has been attached to the relation between both the actors of economics system and innovation activities and other fields of economic policy. But the economy is a system, so it is necessary to understand it to relate both partial decisions and overall regulation. Briefly it is necessary a systemic approach (Smith 1991).

4.2.1 Neoclassical Approach

The neoclassical approach, based on the existence of competitive equilibrium with well-known assumptions regarding strict convexity, constant returns, complete market, perfect information, absence of externalities and public goods and so on, has dominated the analysis of macroeconomic policy. In fact, policy interventions derive from analysis of the welfare properties of competitive systems (Smith 1991). If those assumptions don't hold, a market failure occurs, implying that it is not possible to achieve an optimal equilibrium, and the public sector may intervene to overcome such failure.

The rationale of traditional analysis of technology policy is based on market failure arguments and social inadequacy of private incentive mechanism, owing to a combination of asymmetric information and moral hazard under uncertainty. The policy measures are seen in terms of adjustments towards competitive equilibria and this consideration is true for macroeconomic and microeconomic analysis.

In these standard models the government is assumed to know the market failures and how to correct them. The purpose of the policy maker is to maximize a social welfare function under the constraint that individual agents maximize their selfish utility function. The social planner is well informed about the economic situation and is able to intervene efficiently implementing incentive schemes, choosing the level of innovative activity to engage in. Then, R&D is one of those activities that cannot be left entirely to the private sector because this will lead to under-investments. To reverse this trend governments put in place a whole host of especially fiscal measures, ranging from governmental grants and contracts to targeted tax incentives, in order to encourage enterprises to commit more resources to R&D (Mani 2001).

Nevertheless, many researches have demonstrated that the neoclassical approach is unsuited to analysis of problems of technological change (Smith 1991).

The fundamental limitation of neoclassical policy framework derives from the underlying concepts of production and competition. The production is a process of combination of factors and not a technical process of transformation input. Considering that producing and techniques are well not the decisions of firms regarding from one side what to produce and from other side how to produce: two decisions are path independent. Competition instead is a process of optimization of above decisions. In this context, the relevant information is not the technological knowledge but the price information. In this way, the economy is represented like a deterministic system in which the fundamental internal processes are those which adapt to exogenous change rather than those which endogenously produce change (Smith 1991).

The research programme is seen as method of generating the optimal level of information and not as search for new methods of techniques producing endogenous change. The market failure and the consequent government intervention, that is an adjustment towards equilibrium, don't concern internal transformations that may change the nature of competitive equilibrium. So, the policy can be a choice among states characterised by given techniques rather than by processes of search of new states.

4.2.2 Evolutionary Approach

The neoclassical approach starts from the assumption of representative agent, static equilibrium and so on, whereas the evolutionary approach, by using biological metaphors, starts from two assumptions: the consideration of heterogeneous agents or economic units (firms, consumers, or even technologies) disclaiming the standard neoclassical concept of representative agent, and dynamic equilibrium. Evolutionary economics, inspired from Schumpeter's (1912) notion of disequilibrium dynamics resulting from the introduction of innovations, pays particular attention to the role of technology and institutions in the process of economic growth.

Social system is conceptualized "*as compose of different domains e.g. the separated domains of technology, economy and institutions.... Each of domains has its own dynamics and explanatory process, but what is important is that the domains expert strong mutual influence. Thus, the perspective offered by these theories is that of the world economy as a process of constant transformation. Technologies and institutions change over time, and what drives economic growth in one era might become much less important, or might be substituted by a different factor in a different era. In terms of economic growth rates, such a process is quite different from the neo-classical notion of steady-state growth*" (Verspagen 2001, p. 5).

Each system is a set of elements that interact with each other or even in relation to each other; it is also an element of another system. This explains, why when speaking of interaction or interdependence, we also speak of mutual influence of the elements. Any shift or change of a component of an organization (as a system) makes a change of the system itself in its totality (Barile et al. 2012). Under those assumptions,[3] the systems theory observes entities and their environment through a systemic viewpoint, starting with the analysis of fundamental elements and finally considering more complex related systems. Each system is related to other systems, placed at higher level of observation, defined supra-systems, whose traits can be detected in their own subsystems.[4] The analysis concerns a system made up of many parts or structures. In this sense, every entity (a firm, or simply an individual, a consumer, or a community) as a system can be considered a micro-environment, made up of a group of interlinked sub-components which aim towards a common goal.

[3]See also Barile (2000, 2008, 2009), Golinelli (2000, 2005, 2010), Barile et al. (2012).

[4]Principle of system hierarchy.

In this context, the policy based on the repair of a defective co-ordination mechanism (market failure) has not relevance, but policy interventions are related to shaping the potential technological trajectory.

Evolutionary approach concerns endogenous change, evolution and economic development, so that the policy question is that of increasing creativity, technological opportunity and market development.

The policy maker is no more considered as a fully informed economic agent, having a better understanding of market situation and technological knowledge. On the contrary, he has to learn about the different situations and about the policies he has implemented in order to adapt them in case of inefficiency. The objective of a policy is generally not to reach a predetermined result or technological output, but to improve innovation processes, learning abilities and adaptive behaviours of economic actors and interaction between them, to foster the competitive performances of the economic actors and systems.

The policy has then to stimulate the technological and innovative capabilities of organizations, where stimulating innovation means enhancing learning processes of organizations, generating and coordinating variety within the economic system, influencing the various selection mechanisms in order to create new knowledge (Llerena and Matt 1999).

There is a need for policy intervention to improve the performance of the system by coping with the technological evolution. The modes of intervention cover a large range of policies, from education to technology policies, from generic R&D expenditures incentives to public procurements. In the evolutionary context, the scope for policy is not optimise with respect to some objective function but rather to stimulate the introduction and spread of improvements in technology (Nelson and Nelson 2002).

There are some central issues in technology policy as seen from evolutionary perspective. A crucial distinction between policies which take the innovation possibilities of firms as given and those which seek to reduce the cost of research to the firm. R&D subsidies and tax incentives for R&D are typical example or policies to increase the pay-off to innovation either in terms of public procurement of R&D intensive products or through the duration and scope of patent protection. Policies to change the innovation possibilities of firms would include collaborative R&D programme and policies to ink internal efforts of firms with public R&D carried out in the science base.

In an evolutionary framework, policy should be structured around important arguments: operation of economic and social mechanism which generate experiment and search generally; the operation of selection mechanisms at differential levels; the interaction between institutional structures and strategic behaviour. In the evolutionary context, it is the technological performance to become a fundamental object of policy. The aim of innovation policies is to relate to other policy arena, to regulate the differential impacts of other policy actions, to develop and to maintain the ability of firms to operate in technological trajectories.[5] In the last

[5]For a firm the trajectory is determined by large-scale, often global, trends in demand and technological opportunity which are usually uncertain and at best involve risk for the firms.

cases in particular the ability of firms to regain trajectory may depend on policy support.

4.3 The Innovation as Strategy of Economic Growth

It is undisputed that innovative activity has been the most important component of economic growth. Empirical researches[6] bring out that technological change is central in explaining the inter-economy growth rate differences and that inter-industry R&D differences are highly correlated with productivity and growth rate differences. In fact much of public expenditure in R&D is aimed to improve growth performance.

To increase the output of the economy it's possible to choose two ways: by increasing the number of inputs that go into the productive process; by considering new ways to get more output from the same number of inputs by productivity increase. In this fundamental sense innovation drives economic growth. This is one of the most consistent findings in macroeconomics, and it's been true for centuries. Nowadays, policymakers face the economic challenge of competing in a global economy in the midst of a slow recovery from crisis. Solutions could be to increase innovative capacity across the economy so that innovation drives productivity growth and drives economic prosperity and global competitiveness.

4.3.1 The Innovation in Economic Growth Modelling: An Overview

In the 1940s the economist Joseph Schumpeter asserted that "*competition from the new commodity, the new technology, the new source of supply, the new type of organization, competition which commands a decisive cost or quality advantage and which strikes not at the margins of the profits and the outputs of the existing firms but at their foundations and their very lives*" (Schumpeter 1942) assigning a central role to innovation as a driver of growth.

Starting from Schumpeter's descriptive analysis, the importance of technical progress was recognized in the neoclassical growth models (Solow 1956) but the determinants of the level of technology were not discussed in detail and the technology was considered as an exogenous factor. Later Solow developed a formal neoclassical model of growth, based on the concepts of production function where output is a function of inputs and reaches long run equilibrium. Economic growth is the result of the accumulation of labour, capital, and other production factors with diminishing returns to scale. The economy converges to steady state equilibrium

[6]See Innovation and Research Strategy for Growth BIS economic paper n.15 dic 2011.

where the level of per capita income is determined by savings and investment, depreciation, and population growth, but where there is no permanent income growth. In the long run, growth in per capita output depends only on the rate of technological progress. The theory offered no account of how this happened: technological improvements emerge from outside the economic system, and they are not the result of decisions within it.

It is not a theory of the rate of economic growth rather a theory of the properties of an aggregate economy growing at an exogenous rate. Essentially this theory had been constructed to explain various stylised facts of the growth process including the constancy over time of the distribution of income and the productivity of capital (Metcalfe et al. 2000).

From the 1980s, growth research has focused on understanding and endogenizing technical progress, by building into the models knowledge-creating investment, to analyse deeply the sources of long-run growth.

This allowed mutual cause-effect relation between growth and innovation. In endogenous innovation models technological progress is the key to long-run growth, inside the economic process, and it depends on investment in innovation, primarily through investment in R&D and human capital. Increasing returns to scale, which follow from the externality aspects of technological change, is the process used to explain economic growth.

Several models analysed a specific research sector of the economy, which produces both specific new inputs, technical and scientific knowledge. In these models, growth depends both from on increases in the productivity of tools and equipment, resulting from technological change, and from "spillovers" of knowledge among different areas. Because production functions are not independent, and the knowledge input can enter into all firm-level production functions, the spillovers generate increasing returns. In this type of growth models, despite neo-classical growth theory, the growth rate can be permanently increased by activities enhancing the use and the flow of collective knowledge in the system (Romer 1990; Aghion 2005; Aghion and Howitt 2009).

Modern growth theory is largely built on models with constant or increasing returns to reproducible factors as a result of the accumulation of knowledge. New or improved technology can be achieved through its own research and innovations, through the absorption and adaptation of foreign technologies, or through a mix of the two ones.

Innovation is conceptualized as a knowledge production activity. "*Similar to any kind of production activity, inputs are transformed into output where inputs are mainly knowledge and research equipment and output are new products and production processes. This linear view on innovation is well suited for econometric analysis once inputs are proxied by R&D expenditures and outputs by patents. Using such a knowledge-production-function, one can measure the return to R&D investments at regional or national levels as well as the extent to which regions/countries benefit from R&D invested by other regions/countries, otherwise known as spillovers*" (Frenken 2016, p. 3).

Under the assumptions of the linear model (Arrow 1962), the economic question holds whether the R&D investments are below the socially optimal level. As seen briefly above in growth modelling, innovation issue has been a continuing problem, so that on the one side there are some growth models putting the main emphasis on accumulation of (tangible) capital through investment and the growth of the labour force and left all other influences to be subsumed in a "residual factor", on other side, the so-called "New Growth" Theory[7] breaks away from this tradition and moves "intangible investment" in education, research and development to the centre of economic growth[8] and development. The evolutionary approach outlines the importance of a new concept: complementarity.[9]

4.3.2 The Evolutionary Approach to Innovation in Economic Growth

Evolutionary theory criticises neoclassical theory for neglecting the question of the determinants of technological progress. In the neoclassical growth model, the contribution of a new technology to economic growth cannot be discussed for two reasons. First, we usually are referring to a one-sector model in which a homogenous product is consumed and invested simultaneously. Second, technological progress is assumed to be exogenous. In contrast to this, evolutionary theory underlines that technological change is a complex, interacting process of invention, innovation and diffusion.

The evolutionary approach suggests that a useful description of economic systems has to incorporate the emergence of new elements of the economic process - technical and social innovations as well as new knowledge. The growth and the process of economic transformation have to be analysed as evolutionary technologies associated with increasingly higher levels of worker productivity, and the ability to produce new or improved goods and services.

This implies, first of all, differences in economic growth (both over time and between countries) which are difficult to predict ex ante, but often have clear

[7]New Growth Theory (NGT) is the view that technological change is essentially an economic phenomenon, or at least explicable in economic terms. Furthermore, the mechanics of economic growth emphasized by NOT captures the traditional idea of uneven growth: some sectors generate more economic growth than others, for example through the creation of new knowledge.

[8]In the 1991 the World Bank in our report has established that "it is intangible investment in knowledge accumulation, which is decisive in the economic growth rather than physical capital investment" (Freeman 2002).

[9]Nelson and others have pointed to the complementarity of all these variables. "The contribution of capital accumulation to growth depends not only on its quantity but on its quality, on the direction of investment, on the skills of entrepreneurs and the labour Force in the exploitation of new investment, on the presence (or absence) of social overhead capital and so forth (Freeman 2002).

underlying explicative factors ex post. There could be historical periods of convergence during times when institutions and technological developments allow this, but periods of divergence of economic growth must also be expected. It means that in the long run, economic growth is not a process of general convergence. Any distinction between cyclical variations and trend growth is problematic.

In the neoclassical approach to economic growth the production is treated as a mechanical rather than an organic process and the social and organisational aspects of production are ignored. Of course this theory is not compatible with the idea of growth as transformation since it excludes from consideration the most pervasive of all the stylised facts of economic growth, structural change.

In order to analyse these issues, some economists left the traditional equilibrium oriented path of neoclassic theory and argued that the evolutionary paradigm is more adequate for analysing development processes in the system, characterized by strong uncertainty and dis-equilibrating forces,[10] and is composed of heterogeneous actors.

Under evolutionary perspective (Verspagen 2001), as asserted by new growth theory, technology is a key factor driving economic growth and the changes in growth rates, but what is specific to this approach is the question of how technological change contributes to the variability of trend growth rates.

To answer this question we have to distinguish between radical and incremental innovation. Radical innovations offer new possibilities for long-run changes in the trend rate of economic growth. Radical innovations break up the existing economic structure and dependencies in the economy. This causes changes in the growth rate that are again difficult to predict in a detailed way ex ante. Incremental innovations are linked with the diffusion of the radical innovations throughout the economy, and they depend on the specific historical and institutional context.

The distinction between innovation and imitation is also important. Technology cannot be fully appropriated by the firm that develops an innovation and quickly technological knowledge flows to other firms and other nations. While innovation may lead to divergence among firms or nations, imitation tends to erode differences in technological competencies, and therefore lead to convergence. For those reasons innovation and diffusion together may lead to turbulent growth paths.

Another important aspect of an evolutionary approach of economic growth regards the growth of new industries and the decline of old ones. The notion of radical and incremental innovation is pivotal in this issue. Radical innovations open up new possibilities for economic activities and often create new industries, or drastically revitalise existing industries. Incremental innovation is then one of the driving forces behind the growth of these industries. It happened because the process of economic growth is characterised by structural change. All these considerations are linked to the notion of economic selection. In evolutionary

[10]In short when economics are out of equilibrium they stay out of equilibrium. But they always exhibit order and that order reflects, and might be measured in terms of processes of interaction and the patterns of co-ordination that ensue.

economics, competition is seen as a process that is important in terms of its dynamics, not its long-run tendency. The dynamics of selection drive economic growth.

4.4 Innovation Policy in the Globalization Scenario

> With the fall of Keynesianism in the 1980s and the rise in global competition in the 1990s, innovation policy has become the cornerstone of economic policy in every high-income country or region (Frenken 2016, p. 2).

An intensification of world trade and stronger competition of national economics on internationally mobile production factors that put new challenges to economic policy are closely associated with globalization.

The globalizations of technology can entail one or all of the following issues:

1. The global exploitation of technologies through patents and licenses;
2. The global sourcing of R&D through alliances and joint ventures with foreign companies or universities;
3. The global production of R&D through overseas subsidiaries.

In the last years, it has attempted to distinguish different meanings of the globalisation of technology measuring each of them quantitatively and providing appropriate policy analysis on each dimension, through three main categories (Archibugi 2000):

1. The international exploitation of nationally-produced technology;
2. The global generation of innovations by multinational enterprises;
3. The global technological collaborations.

The key to achieve long run economic growth and welfare is to increase learning. Although the benefits associated to each knowledge-intensive transaction will not be equally distributed among the participating countries, the relevant aim of public policies should be to involve national economic agents in knowledge exchanges.

Policy-makers should support and reinforce (and if necessary initiate) structural change, investing public resources (or providing incentives for private investment) in the technological capabilities which define the new epoch of growth. In promoting innovation policies in the globalisation they have to follow two approaches:

1. The first stresses the importance of spatially bounded (local, regional or national) innovation systems, paying less attention to the differences among neighbouring firms operating in different industries;
2. The second approach stresses instead the role of global factors in innovation systems overlooking location-specific aspects of this process.

Policy makers have to try to identify the relative role of regional, national, sectoral and global factors in shaping innovation systems. The innovation systems concept is itself flexible enough to allow us to take into account the relative importance of each of these factors.

4.4.1 National Innovation Systems

For a long time, innovation has been analysed by a linear model. The development, production and marketing of new technology were assumed to follow a well-defined time sequence which doesn't correspond to the real facts of the innovation process. The criticism has led to a broader view of the process of innovation as an interactive process. This approach emphasizes the central role of feedback between phases of innovation and numerous interactions among science, technology and economy. All these assumptions are the foundation of systemic approach for the analysis of innovation processes. According to this approach innovations are seen as part of larger process of knowledge production of economic relevance. The approach stresses that firms do not innovate in isolation, but in interaction with other organizations, other firms, R&D institutions, universities and other forms of producer services (Fisher 2002).

The work in institutional economies has emphasised that knowledge storing and learning activities are heavily dependent on the institutional context. This leads to a broad analysis of the combined technical and institutional learning processes in the dynamics of knowledge.

Analysis of technology performance and policies has traditionally focused on inputs and outputs, the measurement of which is standardized across OECD countries. The limitations of this approach have become evident over time because these indicators, important sources of information about the content and direction of technological endeavour, do not offer convincing explanations of relation between innovation and growth. They do not consider the interactions among various actors in innovation process. Evolutionary theory highlights the role of those interactions among the people and institutions involved in technology development.

According to the systemic and evolutionary perspective the analysing innovative activity is equivalent to consider it as a structure formed by economic, technological and social elements, in relation to each other, and by interacting influence the creation of innovation processes. The innovation is observed as a complex phenomenon, emerging from synergistic interactions between multiple actors and the final resultant is greater than the sum of the actions performed by every single part.

National Innovation System (NIS)[11] is a concept based on the assumption that understanding the linkages among actors involved in innovation process is the way

[11]"The origins of the systems concept, applied to innovation, lie in the concept of national systems of innovation (Freeman 1987; Nelson 1988; Ludvall 1992). The concept emerged as an alternative way to explain the innovation process, improving on an earlier view that considered this process as a simple linear progression of scientific research" (Iizuka 2013, p. 2).

to increase technology performance. It follows systemic and evolutionary approaches to the study of technology development as opposed answer to the "linear model of innovation".[12]

The innovation system is therefore composed of many relevant factors as institutions, political processes, public research facilities (universities, research institutes, public sources subsidies, incentives, etc.), financial institutions, and so on. In this view the dynamics of innovation are explored in their different phases in order to investigate how they influence and are influenced by the social, institutional and economic context composing the structure of the NIS. The point of view adopted in the analysis of a national innovation system is innovative firm as the main responsible of innovative activity. The main organizations with which innovative company interacts are subsystems of the national innovation system. In particular, these other companies (competitors or suppliers) are universities and other scientific research centres, financial institutions and the government (Barile et al. 2012).

The innovative performance of a country depends mainly on how these actors interact to each other as elements of a complex system of knowledge creation and on technologies they use. How technological change and potential growth are effectively reached is the result of how agents can exploit their interactions, their environment and can learn from the past experiences. Therefore in the definition and implementation of innovation policies a strategic role is played by National Systems of Innovation as a set of organizations, institutions, and linkages for the generation diffusion, and application of scientific and technological knowledge operating in a specific country. In the past, the concept of national system had a well-defined meaning when basic decisions concerning the science, technology, and innovation policies of a given country were taken essentially at a national level.

Economic globalization has established a new, hierarchical system which can override actions at the community level. Market integration and instantaneous global communication create the potential for huge economic and social impacts. In the globalisation where the international linkages are dominant the NSIs are becoming more open systems. Moreover, the national borderline is now less meaningful because national power is flowing partly upwards towards supranational institutions and partly downwards towards regional and local institutions.

The evaluation of national innovation systems regards four types of knowledge or information flows: interactions among enterprises; interactions among enterprises, universities and public research institutes; diffusion of knowledge and technology to enterprises; personnel mobility. It can also be analysed at sub-regional,

[12]"Innovation is thus the result of a complex interaction between various actors and institutions. Technical change does not occur in a perfectly linear sequence, but through feedback loops within this system. In the centre of this system are the firms, the way they organise production and innovation and the channels by which they gain access to external sources of knowledge. These sources might be other firms, public and private research institutes, universities or transfer institutions—regional, national or international. Here, the innovative firm is seen as operating within a complex network of co-operating and competing firms and other institutions, building on a range of joint ventures and close linkages with suppliers and customers" (OECD 1997, p. 12).

national, and international levels. Collaborations and international technology flows have growing importance, but national level seems the most significant for the role of country specific interactions in building an innovation climate.

4.4.2 Further Developments of Systemic and Evolutionary Approach to Innovation

Under the same assumptions of National Innovation Systems theory, it has been formulated the "Triple Helix Model" (Etzkowitz and Leydersdoff 2000).

The model is a metaphor which highlights as the power of the innovative capacity of a country system or of an innovative local system depends on the synergistic interactions among the three main institutions (or "helix"):

1. The state (national and local public institutions);
2. The research system (universities and research organizations);
3. The production system (companies with high technological intensity).

This model tries to capture the reciprocal interactions between the actors on the corporate knowledge process. In the Triple Helix model the university plays a key role bringing its action at the level of other spheres and adding to the exclusive role as producer of knowledge and the subject directly active in promoting innovation. Finally, the State and, more generally, government institutions take on a new and more modern function. In the Triple Helix model, government institutions, whether central or regional and local, not only perform the task of supporting financially the research activities of other entities by public funds.

The action of the three subjects is observed as a sort of "Triple Helix" in which the interactive relationships among the spheres are constantly created. The spheres act in complementary and continuous actions, almost playing each other's role, not losing the sight of their mission.

In this way, the state is committed to defining new rules that favour a free search of high quality, which promote a permanent and stable co-operation between universities and enterprises. This encourages qualified research personnel training, which favour the mobility of researchers between the public and the private sectors, the valorisation of research results, the creation of new businesses from university research activities and the spread of the venture capital business.

The state becomes the subject who writes the rules and ensures the respect by systemic relationships between businesses and universities. The goal is to realize the best framework and system conditions to promote the growth of the innovation capability of the country.

The systems approach, as argued (Barile et al. 2012), is one of the most important recent revolutions for the study of complex phenomena such as the innovative activity. However, none of the theories set forth above is limitless. The Theory of National Innovation Systems, while highlighting the systemic nature of innovative activity becomes less and less sustainable in the current context characterized by increasing levels of globalization and dematerialization that reduce the time-space boundaries. The Triple Helix model has the merit of raising the role of universities in economic entity which, turning research into business ventures is able to transfer knowledge to the market and contribute to the development of the system. However, some academics have indeed expressed some concerns. They fear that the main missions of the universities (teaching and research) may be penalized. However, it was observed that a greater integration between the academic world and the business world is without prejudice to research and training and contributes to increase the transfer flow of knowledge to science and technology, to stimulate the creation of new jobs and new forms of work and to increase the areas and forms of funding of university research.

Acting on the research capabilities or on the social and human capital reinforces the new knowledge creators, acting on the absorption capacity or the technological and innovative performance, strengthens the diffusion and transfer of technology. For this process, are important not only human resources, but also the stock of human capital and the development of new skills and knowledge that can ensure the constant cultural and professional development (Barile et al. 2012).

4.5 Policy Implications

The theoretical assumptions of NIS are the evolutionary view of technological innovation and economic growth.[13] The concept emerges with the aim to increase policy capacity for national economic growth, in response to the challenges raised from new technologies and international competition. The NIS approach is developed for policy innovation that consider change, complexity and systemic approach. Specific models are characterized by interactions between institutions and technology within individual systems, in which elementary actors create change and adapt to change to mitigate tension and disorder that arise from change.

From evolutionary perspective the policy maker has to consider a considerable deal of indeterminacy and uncertainty. Traditionally a policy maker for reaching equilibrium is obliged to correct the system deterministically with his maximizing calculation. In the NIS approach a policy maker is a part of the systems and due to evolution he must adjust the policy over time. This mechanism of adaptive policy is in contrast to conventional thought. Also the role played by a policy maker is seen differently from conventions as he stands in close interactions with the systems which he coordinate. A policy maker must be adaptive, adjusting a policy with the operation of the system to which the policy is aimed to introduce an innovation".

For all those reasons, a national innovation system offers new insights for government technology policies.

[13]For a deep overview see Shulin (1999).

> Most government intervention in the technology area has been the aim to correcting market failure. In the interest of maximising returns to the general public, technology policies have focused on stimulating or supporting R&D spending by industry through instruments such as R&D tax credits and subsidies. The concept of national innovation systems directs the attention of policy makers to possible systemic failures which may impede the innovative performance of industry. The lack of interaction between the actors in the system, mismatches between basic research in the public sector and more applied research in industry, malfunctioning of technology transfer institutions, and information and absorptive deficiencies on the part of enterprises may all contribute to poor innovative performance in a country (OECD 1997, p. 41).

NIS studies (OECD 1997; Shulin 1999) consider the need of new types of policies to overcome systemic failures, and in particular policies addressed to networking. Networking schemes improve the interaction of actors and the role of institutions within national innovation systems. Such policies put the emphasis on the role of joint research activities and other technical collaboration among enterprises and with public sector institutions; policies aimed to promote research and advanced technology partnerships with government are important in this context.

It has been distinguished (Metcalfe 1995) also technological policies of political parties, especially of the government, into two categories:

- Policies to create infrastructures that promote innovative activities both regulative that physical, as buildings, transport, telecommunications, science and technology parks (Barile et al. 2012);
- Policies to develop a specific technology through an indirect intervention (subsidies to private enterprises to encourage them to work on that specific technology) or through direct government intervention (creation of public laboratories that work on that specific technology or in research and development).

Technology policies and infrastructures are implemented by informal flows of knowledge and access to technical networks. They see the value of encouraging the development of innovative relations among firms, and thus establish appropriate competition policy frameworks.

Another policy priority is to increase the innovative capability of firms. From the innovation systems perspective, this means to make enterprises able to access the appropriate networks, to find and identify relevant technologies and information, and to adapt such knowledge to their own needs. It may translate in more investment in internal R&D and information technology. Innovation policies should not only regard the capabilities of individual firms but also the networking and innovative performance of firms and sectors that driven economic growth.

4.6 Conclusions

In the globalisation era innovation issue has becoming a much more important question to economic development. At present, we are looking for rapid and radical technological change that underlies some of the key economic developments.

Technological change provides a privileged viewpoint from which to understand the dynamics of globalisation. New technologies have always been international scope; the transmission of knowledge has never respected states' borders.

There is a complex interplay between technological change and globalisation. On the one hand, new technologies act as a powerful vehicle for the diffusion of information across distant communities. For example, it would be difficult to imagine the current globalisation of financial market without the existence of the new information and communication technologies, since they have made it possible to obtain instant transactions across the world. On the other hand, the process of generating and discussing new technologies has been moulded and strengthened by the flows of individuals, commodities and capital.

Globalization represents a change in human system complexity. To deal with it, it is necessary to assure variety of responses and flexibility. We need a creative approach on all the components of the action process: from organization, to problem definition, to solution design. To perform these tasks, complex systems are required to support institutions and good economic policies, in particular a policy co-ordination both between different sector policies and across different territorial governments. This requirement has to be met because innovation policy affects the capability to innovate so it needs to be co-ordinated with policies affecting the transformation pressure and with policies affecting income distribution.

The ability of government in this scenario is to adapt the policies to the different situations.

The adoption of a complexity and evolutionary perspectives to innovation can help policymakers to realize policies able to promote innovation processes. Policymakers need new tools to understand the implications of their policies at all different involved levels. Policy interventions directly influence and determine interactions and changes within sectors, firms or territories. They have a deep impact on macro variables such as imports and exports, investment, overall expenditure both in public and private R&D.

In other words in a systemic/evolutionary approach innovation policies should be evaluated with respect to the systemic effects they produce.

This implies that policies should be monitored and evaluated in new way, by devising new indicators, new procedures that consider either the various needs that different innovation processes require in order to produce tangible results in term of economic growth, either the processes of interaction among the involved actors.

For example, while countries can stimulate investments in R&D in private sector by merely fine tuning financial instruments, such as research grants and tax incentives in the case of developing countries a mere fine tuning of the financial instruments while necessary are not sufficient. An important element or technology policy has to be the development co-operative strategies aimed at collective provision of scientific research. In other words, policy might well be proactive—and should often be. In other words, a 'problem' might be something that has not yet emerged.

Central questions of policy concern how do firms and governments develop strategies for science, technology and innovation; which are the inter-relations between public and private in the creation of technology; which is the proper role of the public sector in this area; which are the implications of increasing internationalisation for national science policies.

The policy maker can no longer look to research programmes which aim simply at the development of new scientific and technological principles and results, but it is increasingly necessary to focus also on the application and use of science and technology, by companies and by society as a whole because we are living in globalisation era.

References

Aghion, P. (2005). *Handbook of economic growth theory*. Cambridge, MA: MIT.

Aghion, P., & Howitt, P. (2009). *The economics of growth theory*. Cambridge, MA: MIT.

Archibugi, D. (2000). The globalisation of technology and the European innovation system, Prepared as part of the project "Innovation Policy in a Knowledge-Based Economy" Commissioned by the European Commission.

Arrow, K. J. (1962). The economic implications of learning by doing. *Review of Economic Studies, 29*, 155–173.

Barile, S. (Ed.). (2000). *Contributi sul pensiero sistemico*. Salerno: Arnia.

Barile, S. (2008). *L'impresa come Sistema – Contributi sull'Approccio Sistemico Vitale* (II ed.). Torino: Giappichelli.

Barile, S. (2009). *Management Sistemico Vitale*. Torino: Giappichelli.

Barile, S., Polese, F., & Saviano, M. (2012). *Immaginare L'Innovazione*. Giappichelli Ed, Torino.

BIS-Department for Business Innovation and Skills. (2011). *Innovation and research strategy for growth*. BIS Economics Paper, no. 15. http://bis.gov.uk/innovatingforgrowth

Etzkowitz, H., & Leydersdoff, L. (2000). The dynamics of innovation: From national systems and "Mode 2" to a triple helix of University-Industry-Government relations. *Research Policy, 29*, 109–123.

Fisher, M. M. (2002). *A systemic approach to innovation in regional development reconsidered Part A* (pp. 15–31). Berlin: Springer.

Freeman, C. (1987). *Technology and economic performance: Lessons from Japan*. London: Pinter.

Freeman, C. (2002). Continental, national and sub-national innovation systems-complementarities and economic growth. *Research Policy, 31*, 191–211.

Frenken, K. (2016). *A complexity-theoretic perspective on innovation policy* (ISU Working Paper #16.01).

Golinelli, G. M. (2000). *L'approccio Sistemico al Governo dell'Impresa – L'Impresa Sistema Vitale* (I ed.). Padova: CEDAM.

Golinelli, G. M. (2005). *L'approccio Sistemico al Governo dell'Impresa. – L'Impresa Sistema Vitale* (II ed.). Padova: CEDAM.

Golinelli, G. M. (2010). *Viable systems approach (VSA). Governing business dynamics*. Padova: Kluwer CEDAM.

Iizuka, M.. (2013). *Innovation systems framework: Still useful in the new global context?* (UNU-MERIT Working Papers 2013-005).

Kline, S., & Rosenberg, N. (1986). An overview of innovation. In Landau et al. (Eds.), *The positive sum strategy*. Washington, DC: National Academy Press.

Lane, D. A., & Maxfield, R. (1997). Foresight complexity and strategy. In W. B. Arthur, S. Durlauf, & D. A. Lane (Eds.), *The economy as an evolving complex system II*. Redwood City, CA: Addison Wesley.
Lane, D. A., & Maxfield, R. (2005). Ontological uncertainty and innovation. *Journal of Evolutionary Economics, 6*, 43–76.
List, F. (1841). *The national system of political economy* (S. Sampson & M.P. Lloyd, Trans., 1885 ed., Fourth Book, "The politics", Chapter 33).
Llerena, P., & Matt, M. (1999). *Technological cooperation and dynamic performance of industries: Towards an evolutionary foundation of technology policies*. Communication to European Meeting on Applied Evolutionary Economics 7–9 June, Grenoble, France.
Ludvall, B.-Å. (Ed.). (1992). *National systems of innovation and interactive learning*. London: Pinter.
Mani, S. (2001). *Role of government in promoting innovation, an international comparative study*. In Paper presented to the Conference "The Future of Innovation Studies Eidhoven University of Technology, The Netherlands 20–23 September.
Metcalfe, J. S. (1994). Evolution, technology, policy and technology management. *Prometheus, 12*, 1.
Metcalfe, S. (1995). The Economic foundations of technology policy: Equilibrium and evolutionary perspectives. In P. Stoneman (Ed.), *Handbook of the economics of innovation and technological change*. Oxford: Blackwell Publishers.
Metcalfe, J. S., Fonseca, M. D., & Ramlogan, R. (2000). *Innovation, growth and competition: ESRC Centre for research on innovation & competition*. In Paper prepared for the Complexity and Complex Systems in Industry Conference 19th 20th September, University of Warwick.
Nelson, R. R. (1988). Institutions supporting technical change in the United States. In G. Dosi, C. Freemann, R. R. Nelson, G. Silverberg, & L. Soete (Eds.), *Technical change and economic theory* (pp. 312–329). London: Pinter.
Nelson, R. R., & Nelson, K. (2002). Technology, institutions, and innovation systems. *Research Policy, 31*, 265–272.
OECD. (1997). *National Innovation Policy*. Paris: OECD. https://www.oecd.org/science/inno/2101733.pdf
Romer, P. (1990). Endogenous technological change. *Journal of Political Economy, 98*, 5.
Schumpeter, J. (1942). *Capitalism, socialism and democracy*. London: Routledge.
Shulin, G. (1999). *Implication of National Innovation Systems for developing countries managing change and complexity in economic development* (UNU/INTECH Discussion Papers ISSN 1564-8370).
Smith K., (1991). Innovation policy in an evolutionary context. In J. S. Metcalfe (Ed.), Evolutionary theories of economic and technological change. London: Harwood.
Solow, M. (1956). A contribution to the theory of economic growth. Robert M. Solow. *The Quarterly Journal of Economics, 70*(1), 65–94.
Verspagen, B. (2001). *Economic growth and technological change: An evolutionary interpretation* (OECD Science, Technology and Industry Working Papers, No. 2001/01). Paris: OECD Publishing.

Chapter 5
Value-Driven Conceptualization of Services in the Smart City: A Layered Approach

Leonard Walletzky, Barbora Buhnova, and Luca Carrubbo

Abstract The concept of the Smart City is becoming increasingly popular. In essence, the added value of a Smart City emerges from the harmony of services that realize the foremost goal of the Smart City, which is high quality of living via improved efficiency of city-relevant processes and citizen engagement.

However, the services that form a Smart City are nowadays considered in isolation, both in theory and in practice. Moreover, services on different levels of abstraction (key services vs. supportive services) are often depicted side by side without understanding that one is the supportive pillar for one another.

This paper identifies the benefits of an integrated view that not only interconnects the services, but also identifies joint layers that they rely on, which helps us to understand the impact of the underlying IT services and the infrastructure they rely on. At the same time, we extend our view to the Smart Citizen, who plays an essential role in the value creation process within the Smart City.

Keywords Smart City • Service research • Smart service systems • Win-win logic

5.1 Introduction

Smart City can be defined as a city seeking to address public issues via the use of information and communication technology (ICT) to sense, analyze and integrate the key information of core systems in the city (European Parliament, Directorate General for Internal Policies 2014; European Commission 2012). However, although the foremost goal of the Smart City is to leverage the value of available city services via their integration, the current state of the art is very far from reaching this goal. Smart City services are being designed, built and deployed in

L. Walletzky (✉) • B. Buhnova
Faculty of Informatics, Masaryk University, Brno, Czech Republic
e-mail: qwalletz@fi.muni.cz; bara.buhnova@fi.muni.cz

L. Carrubbo
SIMAS Interdepartmental Research Centre, Salerno University, Fisciano, Italy
e-mail: lcarrubbo@unisa.it

S. Barile et al. (eds.), *Social Dynamics in a Systems Perspective*, New Economic Windows, DOI 10.1007/978-3-319-61967-5_5

isolation, which has two critical consequences disregarded by the relevant communities.

First, the architects of Smart City services are relying on the unrealistic assumption of guaranteed availability of common infrastructure (whether related to ICT, such as the communication network, or unrelated to ICT, such as the traffic roads). This may become highly problematic once the services are deployed and start to compete for the shared resources.

Second, very low attention is being paid to the human aspect of Smart Cities, and mainly its citizens. It is clear that the citizen engagement (implying the concept of a Smart Citizen) is what effectively determines what effect the smartness of the city has on the actual quality of life, which is the ultimate goal of the transition towards Smart Cities.

Although these two aspects are being to certain extent present in existing Smart City models, the models are very simplistic, revealing very little understanding of the role of these two aspects, and commonly misplacing them in it. We argue that when these two pillars are misplaced in the Smart City architecture, it is very difficult to recognize the dependencies among the Smart City services, which is in our opinion why no dependency matrix for Smart City services can be found in literature. We believe that the conceptualization of the layers in the Smart City service ecosystem is the missing step towards the maturity in this domain.

In this work, we introduce a holistic view of the Smart City architecture that integrates these two perspectives into the big picture. Our approach uses the concepts borrowed from layered software architecting to conceptualize several levels of abstraction in the ecosystem of Smart City services, together with the dependencies among services and their supportive mechanisms (including the software and hardware they rely on). The novel contribution of this work includes, besides the layered view and the dependency service view, also the reasoning about the propagation of service usage to the underlying features, services, ICT infrastructure, and then back to the user, which can help the Smart City architects to identify critical elements of the Smart City architecture and remove the constraints associated with them. We claim that the understanding implied by the recognition of such dependencies and critical elements can help Smart City architects in making better decisions when designing the Smart City.

This chapter is structured as follows. Section 5.2 first reviews the related work. Then Sect. 5.3 details a Smart City analysis, namely from the two discussed perspectives of both a Smart Citizen and ICT infrastructure. Finally, Sect. 5.5 introduces the concept of layering in Smart City service architecture, Sect. 5.6 discusses the findings and Sect. 5.7 concludes the paper.

5.2 Related Work

Numerous models of Smart City services exist, which may appear fairly complete at first glance. However, at second glance, they are all very simplistic when it comes to understanding the dependencies among the services and different supportive role they play for each other.

Our perspective on Smart City architecture has been inspired by two points of view:

- Functional view, which emphasizes the service provision and namely the utility of the services towards the citizens. This approach often takes certain sociological or psychological analysis into account.
- Technical view, which emphasizes the ICT perspective and ICT services that drive also the design of the structure of the discussed Smart City architecture.

A typical example of the models focusing on the functional view of Smart City services is depicted in Fig. 5.1 (Singh, http://www.forbes.com/sites/sarwantsingh/2014/06/19/smart-cities-a-1-5-trillion-market-opportunity).

In this view, there is only one concept layer containing all the services with very unclear dependencies among each other. In the example, Smart Infrastructure (which is deeply technical) is put side by side to Smart Energy (that should strongly rely on Smart Infrastructure) and even Smart Citizen (who we expect to rely on almost all other smart services). In general, Smart Infrastructure is apparently important for every other service whereas Smart Citizen is mostly a prediction of functionality and not a kind of a service. It is very confusing to analyze a Smart City while relying on such a simplistic view.

Fig. 5.1 Smart City structure

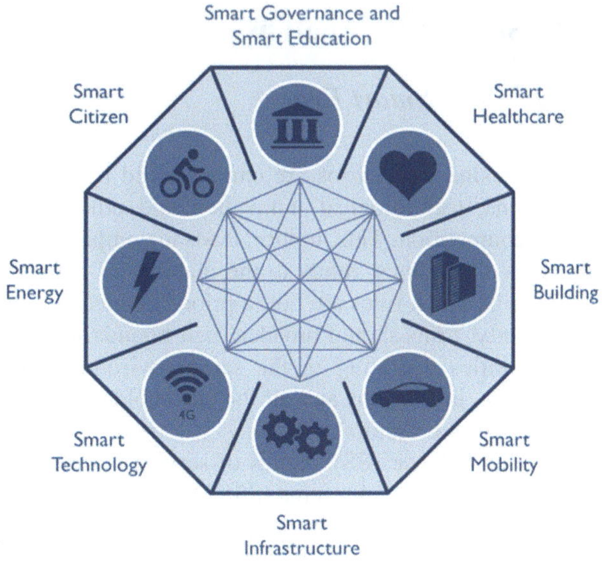

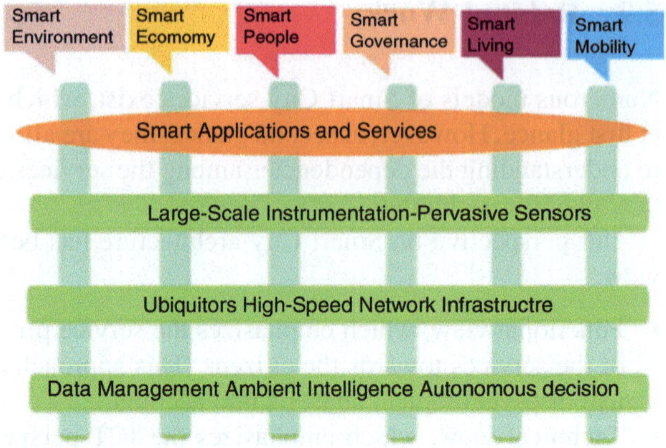

Fig. 5.2 The Smart City technological structure

The second point of view is demonstrated in Fig. 5.2 (Balakrishna 2012), where the ICT perspective is detailed, but without any other consequences or deeper analysis. To be able to describe Smart City in its complexity, it is necessary to combine both these concepts not only together, but also with other disciplines.

The model of the Smart City can be seen as a kind of ecosystem. As shown in (Caputo et al. 2016), Smart City (or Community) can be described within the conceptual model of Social, Business and Digital Ecosystem. Therefore, the model of the Smart City should contain many different aspects and views—the utility for the citizens as well as the description of the way how the particular service is provided and what inputs it requires for proper provision.

5.3 Smart City Analysis

5.3.1 New Smart Era

Today business scenarios are characterized by globalization, social and political evolutions, technological innovation and other factors that determine a rising hyper-competition. This, in turn, leads to complexity. This phenomenon, examined by many disciplines, refers to the inability of organizations to act in an unstable context, in which rules are not 'a priori' defined and the risk is high. To tackle this uncertainty, firms should broaden their boundaries and establish relationships with other entities operating in the same environment (i.e., other public or private companies, institutions and all the stakeholders involved in their conduct; we can refer to these as Actors).

Service Science (SS) studies of service systems and particularly the advances about Smart Service Systems (SSS) in any field of interest foster the ground upon which scholars and practitioners try to deepen new ways of decision-making in order to reduce surrounding complexity.

The evolution of the Service Research has proposed the development of a study, which focuses on modern service systems, intelligent, smart, driven in particular by the progress made at international level in ICT. The idea is based on the need to consider more organizations better able to deal with changing conditions in a more responsive, adaptive, proactive, dynamic context. In order to deepen the new reflections on the concept of SMART (acronym of "specific, measurable, agreed, realistic and timely") (IfM, IBM 2008), many researchers have in recent years been investigating all potential service applications, defined "on stage", referring to the practical evidence of something truly iterative, interactive, interconnected, intelligent, and that is a representative of a smarter planet. Considering a real overview, we wanted to understand how to achieve the sustainable development of a complex system, characterized by many actors (workers, citizens, producers, suppliers, authorities, consumers, users, etc.) and many facilitators (retail sales, distribution roads, informative networks, agriculture, financial services, healthcare government), which are fundamental for the improvement of management capacity and implementation of collaborative strategies. In this sense, the intelligence service systems are not derived from intuition or chance, but by systematic methods, continuous learning, data collection, rational innovation, social responsibility and governance network. Applying an intelligent service, practices smart, inserted in smart cities, with intelligent organizations, through intelligent operations, for intelligent results, there may be some major changes in our daily lives.

Service systems are smart if they react to circumstances and make a rational and efficient use of resources, indeed all the sub-systems of any organization share a common goal, a determined finality pursued by the whole system, similar to the way in which smart service systems are oriented towards enduring performance and towards satisfaction of all the involved actors.

The governance of organizations should direct the system towards a final goal, transforming static structural relationships into dynamic interactions with other entities operating within the same service ecosystem (Wieland et al. 2012; Siltaloppi et al. 2016; Polese and Carrubbo 2017). The decision maker, by adopting a systems framework, could be capable to use the continuous information exchange and increases comprehension of the context in which he/she operates, by being able to valorize the multiple contributions of Actors within the process of value generation (Polese et al. 2016).

5.3.2 *The Complexity of Smart City*

The complexity of the Smart City concept calls for the recognition of different axes that can be used to structure the services within the Smart City ecosystem. Such axes can then guide the split of the services into categories and analyze the relationships among them.

Although the most common structure of services follows a plain flat hierarchy of services, viewing the Smart City as a mix of services at the same level (see Fig. 5.1), attempts towards more structured approach can be identified too. For instance, Taewoo and Pardo divide the services into three levels, which are (1) human, (2) technological, and (3) institutional (Nam and Pardo 2011). Such an approach is however not ideal either, as numerous services internally mix different aspects of these three perspectives.

Our strategy to service ecosystem (Vargo et al. 2015; Koskela-Huotari et al. 2016) conceptualization has been highly driven by the recognition of the fact that the services provided by the city or its suppliers must be useful for the citizens and/or other city stakeholders (Vargo and Lusch 2008). In this sense, it is important to emphasize, that the service provider cannot directly deliver value to the stakeholders, it is only the value proposition that can be delivered. It must be therefore made explicit, what is the value proposition associated with the individual services, and what is the potential value resulting from it.

At the same time, not all the services related to the concept of the Smart City are valuable directly for the citizens. There are numerous services that are playing their role in the scheme of the smart city and that must be included into the big picture, although they only have indirect impact on the life of the citizens. These include for instance the hardware infrastructure, the services of database servers, or document management.

One of the comprehensive lists of Smart City domains representing value for the citizen have been collected by Neirotti et al. (2014) who distinguish: (i) *Energy grids*, which employ ICT to enable advanced load optimization and pricing for the customer; (ii) *Public lightning and natural resources and water management* that focuses on utilization of renewable resources; (iii) *Waste management* which deals with coordinated waste collection, disposal and recycling; (iv) *Environment* for continuous sustainability and pollution control; (v) *Transport and logistics* which offer services related to the traffic control; (vi) *Office and residential buildings* which adopt modern technologies to create efficient living and working environments, (vii) *Healthcare* facilitates services for disease diagnostics and healthcare; (viii) Public security that involves technologies employed by police and fire departments for critical incidents detection and prevention. This list became the starting point for our layered taxonomy presented later in this chapter.

Because smart service systems inevitably involve multiple actors, the organizational configurations need to take account of network theory especially the networking forces and enablers required to keep the system tight and focused towards its goals.

In contemporary research into 'smart service systems', network studies are playing an increasingly important role including studies of resource allocation and the advantages of collaboration, alliances and cooperative strategies.

As the world is becoming smarter[1] (we're talking often about smarter planet), to adapt the systems must be people-centric, information-driven, e-oriented, for a

[1]The concept of a 'smarter planet' refers to an interconnected globe in which there is growing attention to data measurement, the development of networks, enhanced learning, and responsive adaptation processes (www.ibm.com/think).

mutual satisfaction of any Actor involved within, and community should encourage and cultivate people to collaborate and innovate (Qiu et al. 2007).

5.3.3 Value as Deciding Factor

The concept of the model introduced in this chapter is based on the functional logic, used in the domain of information systems design. The main purpose of the model is to separate the functional services that are bringing the value to the citizens from the other services (which are the being further divided into layers). This layer of the services that have their value proposition formulated directly for the citizen, is in our approach called Smart Features (of the Smart City). The smart features include for example Transport, Energy or Healthcare.

The value concept from this point of view certainly follows the perceptual logic. If a value proposition is preferred to another through the mechanism of purchase or the use (value in use) (Vargo and Lusch 2008), it triggers a process of value co-creation (Prahalad and Ramanswamy 2004; Ballantyne and Varey 2006), which implies that the active multi-Actor participation by all the protagonists of the exchange and that makes "effective" value of a particular offer (up to that point only "potential"), which is perpetrated though time if the proposal continues to be attractive to the user (Barile et al. 2013).

Attempting to interpret these interesting conceptual intersections in systems, values, and service, we can deduce that (Carrubbo 2013):

- Value is considered to be an improvement in a system, as perceived by the system itself or by the ability of the system to be integrated in its environment.
- Value creation takes place as a potential resource has become an effective specific benefit.
- Value co-creation has a win-win logic that considers the interaction among different entities represented by various service systems and by the desire to reach collective mutual satisfaction, in which the active contribution is multiple, the integration is maximum, and complementarity is fundamental.
- The contributions of knowledge, the application of skills, the ability to configure and reconfigure, and the desire to maintain relationships with long-terms subjects considered strategic all represent the elements of a systemic way of being adaptive.
- Value is perceived and determined by the customer on the basis of value in use (through the previously defined consumption process); much more than something as defined ex-post, but achievable especially ex-ante, through the relevant contribution recipients in the value co-creation process.

  Then the value co-creation logic is defined in such a sense of win-win, considering the interaction between different entities represented by various Service Systems and the desire to gain a collective mutual satisfaction, in which the active contribution is multiple, integration is maximum,

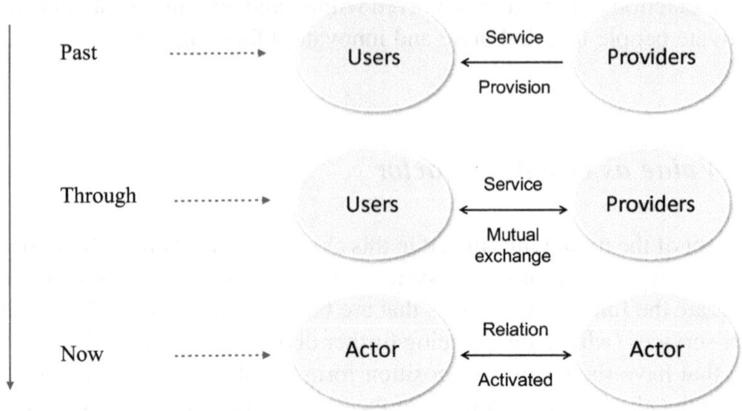

Fig. 5.3 Win-win logic evolution

complementarity is essential. The "win-win" interactions develop only through the promotion and maintenance of relationships with stakeholders or through a common wants to encourage the process of co-creation (not-opportunistic behavior, long-term relationships and shared values) (see Fig. 5.3).

5.3.4 Smart Features as a Gateway to the Smart City Architecture

Smart Features are complex services that are being realized with a complex of more specific services. For instance, the Transport smart feature can be supported with Electric mobility, Traffic control or Public transport services. Those specific services may have more than one value propositions, because they can bring a value not only to different Smart Features (for instance Electric mobility belongs both under Mobility and under Energy), but also directly for the city residents who can use them separately from the Smart Features. For this purpose, they will have unique value proposition formulated in the same way as Smart Features do. Services with more than one value proposition are called Smart Services.

When we go deeper into the structure of the elements of the value proposition and explore the process of the value creation, we are able to find another level of the supportive services, without which we could not be able to establish any kind of the smart service or feature. They are still very complex, but it is quite hard or even impossible to formulate the value proposition for the citizens if they would work in isolation from the other services.

This is the reason why we call them Supportive Services. As an example we can name Traffic monitoring, Charging services, Route planning, or Smart Grid maintenance.

If we analyze the way how the supportive services are formulating their value proposition, we can still see very complex and unique environment of IT systems services.

5.4 Supportive IT Services

Information and communication technology (ICT) plays an essential role in the Smart City to provide support for the Smart Services. The role of ICT is however not very well understood, which results into isolated ICT support for each service individually without understanding mutual interdependencies among them (whether beneficial or possibly harmful).

In principle, Smart City services have very much in common. In effect, they are often built from very similar ICT software elements (discussed below), and rely on a shared ICT physical HW infrastructure. That means that they either compete for the infrastructure or the infrastructure is duplicated. The first option of shared infrastructure is more beneficial from different viewpoints, but we need to understand the mutual impacts of the Smart Services operating on the shared infrastructure, so that negative turbulences are avoided. This is why a holistic understanding of this domain is so beneficial.

From the software perspective, the following key ICT supportive services (referred to as Software IT Services) can be identified, that are necessary for most Smart Services to operate:

- Data collection—the service of gathering different types of data from different sources (devices, sensors, etc.).
- Storage—the service enabling to store the selected data for their later use and processing.
- Event processing—online and offline processing of events necessary for later decision support.
- Control—commanding of devices in the infrastructure (for instance to switch on/off or to update their software).

Through this layer of Software IT Services and its value propositions we can identify the last layer in our architectural view, which is the hardware layer. The hardware layer consists of the elements of the physical infrastructure supporting the layers above, which typically include:

- Sensors—to measure and collect information about different devices.
- Actuators—to command devices when urged by control mechanisms.
- Servers—to store data and run computation and processing.
- Network—to mediate communication among other elements of the infrastructure.

5.5 Concept of Layers

This section presents an example of a semi-populated model of the layers, including some relationships among the services. Although this model is only an initial draft created for demonstration purposes, it is already now visible how complex the whole ecosystem of services in the Smart City is, especially when the interdependencies among the services are visualized (Fig. 5.4).

The upper level brings value directly to the city residents. The service provider is directly the city (or city-designated company) and the target depends on the particular situation of resident's life. The value proposition is based on the residents' needs, not on the possibilities of IT or service design experts. This fact is very important when separating the services (or features) from the others.

The opportunity to explore the processes of a long term interaction in a B2B/B2C/C2C/A2A relationship, as well as the structure of a dynamic system and the expectations of users, identify the 'complexity of the ecosystem' (Basole and Rouse 2008) within which everything is collected, identified and active; such complexity does not only depend on the number of actors, but also by the conditional probability that these actors are involved in the provision of services (Barile and Polese 2010).

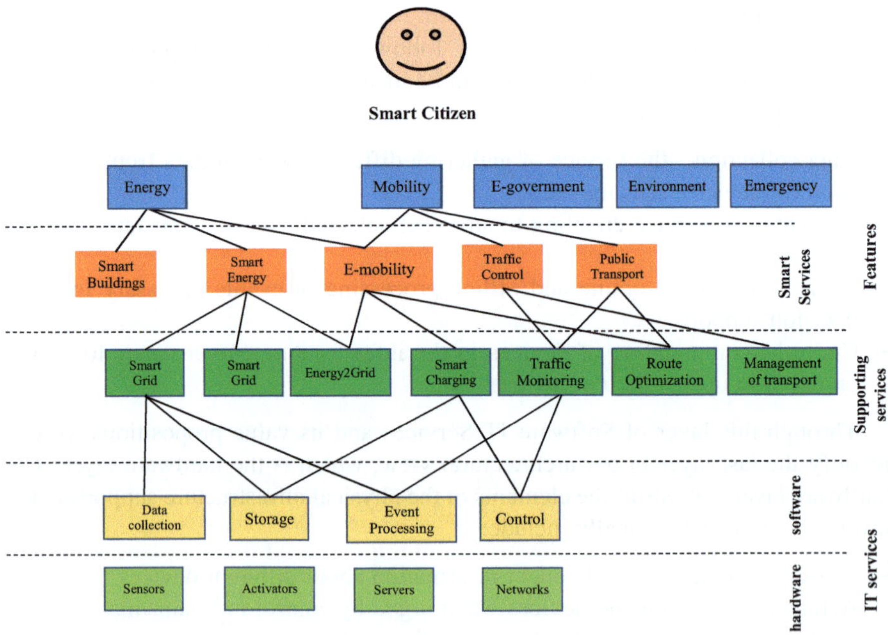

Fig. 5.4 The structure of Smart City layers

The system in this regard is made viable by the behavior assumed (including the perspective of value creation), more strategic, more responsive, more adaptive, more intelligent. The characterization and optimization of the relations, the re-designing of organizational configurations, the management of complexity, are therefore all elements that identify a successful system. In addition, given the perspective of the modern work service oriented, a service system (whose leverages are synergies reticular and co-creation advantages) can be considered as smart if really able to survive within an ecosystem so complex.

The smart services are designed for the best provision of the upper services. But the character of value they bring is different than above. The value proposition is based on improving the support for the services, provided for the residents—faster response, better analysis tool or optimal way of searching. We can find two common signs of second layer services:

- Residents have no or very limited motivation to use them directly (no matter if they know or understand them).
- The providing of the features of the first layer is not possible or is harder without them.

The layer of the supportive services is designed to fit to the needs of Smart Services. One particular service of this layer can support more Smart Services—like Route Optimization is used by Electric mobility (E-mobility) service and also in Public Transport.

The same situation is repeated in the layer of IT services. First in the software part—where we can see each kind of information system is used in more supportive services. And the same logic is used on hardware level—therefore we do not add the relation lines.

5.6 Discussion and Findings

The reasoning employed in this thesis revealed numerous findings about the Smart City services and their interplay that form the Smart City architecture presented in this work. The findings can be summarized as follows.

ICT Services First, we argue that ICT services employed in the Smart City context should not be treated side by side to other Smart City services. It is much more appropriate to consider them as the supportive layer in the Smart City architecture that the higher-level Smart City services are built upon.

Software and Hardware ICT Layer There are in principle two types of ICT supportive services, structured into a software layer and hardware layer. Although the software and hardware elements might be unique for each higher-level Smart City service, the nature of these supportive elements (i.e. sensors, servers, data collection, monitoring) is very similar across different smart services.

Smart Citizen Although some approaches list the Smart Citizen as one of the services, we have identified that this concept rather encapsulate the understanding of the value that is being created via utilization of other smart services.

Value Creation The focus of value creation, and especially of value co-creation, must in this sense be examined both internally (through strategies of improving the quality of goods and services and through structural growth in terms of capabilities, knowledge, and opportunities), and externally (because of the collaborative relationships with other stakeholders).

The value proposition and creation spans across different layers in our layered Smart City architecture, with stronger reflection of this concept in upper levels. Role of the value for the city residents was already identified, but we have identified several services with two different value propositions—for depending Smart Features and for the city residents. Those services are forming sub-layer within the Smart Features layer.

Dependency Analysis Complex "is supported by" dependencies can be identified among the services on different levels of the Smart City architecture. Interestingly, our analysis revealed that the dependencies form a complex graph with a number of services laying in an intersection of different Smart City features (being employed by different Smart City features).

Layering The designed concept of layering is very similar to what is being employed in the standard Software Architecture field of Software Engineering, which means that various advanced techniques of this domain could be transferable to Smart City architecture too.

5.7 Conclusion

In this work, we have introduced an integrated view on the layered architecture of the Smart City, which expands to two directions comparing to related work, i.e. to the human and high-level feature aspects of the Smart City on one hand, which are critical to the value creation process, and to the ICT infrastructure on the other hand.

This extension together with a detailed dependency graph among services on different layers sets the basis for understanding the mutual influence that different layers have on each other, and hence support informed decision on the investments into the Smart City design, as it may help the relevant stakeholders to decide about the critical elements of the city which impact the value creation in the city in the strongest fashion.

The context of Smart Cities is highly complex and brings a number of challenges to the realization of the principles identified in this work.

Extensive further analysis is needed in many directions. In the future, we would like to look deeper into all different Smart Features and decompose them into Smart

Services and their realization via the Supportive ICT Services, so that a complete Smart City architecture can be visualized. Furthermore, the principle of value co-creation among the related services will be examined, as well as the process of value proposition.

References

Balakrishna, C. (2012). Enabling technologies for smart city services and applications. In *6th International Conference on Next Generation Mobile Applications, Services and Technologies (NGMAST)*, 2012. IEEE, 2012.

Ballantyne, D., & Varey, R. J. (2006). Creating value-in-use through marketing interaction: The exchange logic of relating, communicating and knowing. *Marketing Theory, 6*(3), 335–348.

Barile, S., & Polese, F. (2010). Smart service systems and viable service systems. *Service Science, 2*(1/2), 21–40.

Barile, S., Carrubbo, L., Iandolo, F., & Caputo, F. (2013). From 'EGO' to 'ECO' in B2B relationships. *Journal of Business Market and Management, 6*(4), 228–253.

Basole, R. C., & Rouse, W. B. (2008). Complexity of service value networks: Conceptualization and empirical investigation. *IBM Systems Journal, 47*(1), 53–70.

Caputo, F., Formisano, V., Buhnova, B., & Walletzky, L. (2016). Beyond the digital ecosystems view: Insights from smart communities. In *Proceedings of the 9th Annual EUROMED Academy of Business (EMAB) Conference on Innovation, Entrepreneurship and Digital Ecosys.* University of Warsaw, Poland, September 14–16.

Carrubbo, L. (2013). La Co-creazione di valore nelle destinazioni turistiche, ed. RIREA, Collana "Opera Prima", 2.

European Commission. (2012). *ICT concepts for optimization of mobility in Smart Cities.* Publication number 30-CE-0466883/00-79, SMART 2011/0067. December 2012. ISBN: 978-92-79-28716-9.

European Parliament, Directorate General for Internal Policies. (2014). *Mapping Smart Cities in the EU.* Publication number IP/A/ITRE/ST/2013-02, January 2014. Available online at http://www.europarl.europa.eu/

IfM, IBM. (2008). *Succeeding through service innovation: A service perspective for education, research, business and government.* Cambridge: University of Cambridge, Institute for Manufacturing.

Koskela-Huotari, K., Edvardsson, B., Jonas, J. M., Sörhammar, D., & Witell, L. (2016). Innovation in service ecosystems—Breaking, making, and maintaining institutionalized rules of resource integration. *Journal of Business Research, 69*(8), 2964–2971.

Nam, T., & Pardo, T. A. (2011). Conceptualizing smart city with dimensions of technology, people, and institutions. In *Proceedings of the 12th Annual International Digital Government Research Conference: Digital Government Innovation in Challenging Times.* ACM, 2011.

Neirotti, P., De Marco, A., Cagliano, A. C., Mangano, G., & Scorrano, F. (2014). Current trends in Smart City initiatives: Some stylised facts. *Cities, 38*, 25–36.

Polese, F., & Carrubbo, L. (2017). *Eco-sistemi di servizio in sanità.* Torino: Giappichelli Editore.

Polese, F., Tommasetti, A., Vesci, M., Carrubbo, L., & Troisi, O. (2016). Decision-making in smart service systems: A viable systems approach contribution to service science advances. In T. Borangiu, M. Dragoicea, & H. Nóvoa (Eds.), *Exploring services science* (pp. 3–14). Cham: Springer International Publishing.

Prahalad, C. K., & Ramanswamy, V. (2004). *The future of competition: Co-creating unique value with customers.* Cambridge, MA: Harvard University Press.

Qiu, R. G., Fang, Z., Shen, H., & Yu, M. (Eds.). (2007). Towards service science, engineering and practice. *International Journal of Services Operations and Informatics, 2*(2), 103–113.

Siltaloppi, J., Koskela-Huotari, K., & Vargo, S. L. (2016). Institutional complexity as a driver for innovation in service ecosystems. *Service Science, 8*(3), 333–343.
Singh, S. Smart Cities – A $1.5 Trillion market opportunity. Online http://www.forbes.com/sites/sarwantsingh/2014/06/19/smart-cities-a-1-5-trillion-market-opportunity
Vargo, S. L., & Lusch, R. F. (2008). Service-dominant logic: Continuing the evolution. *Journal of the Academy of Marketing Science, 36*(1), 1–10.
Vargo, S. L., Wieland, H., & Akaka, M. A. (2015). Innovation through institutionalization: A service ecosystems perspective. *Industrial Marketing Management, 44*(1), 63–72.
Wieland, H., Polese, F., Vargo, S., & Lusch, R. (2012). Toward a service (eco) systems perspective on value creation. *International Journal of Service Science, Management, Engineering and Technology, 3*(3), 12–24.

Chapter 6
Service as Entropy Reduction: A Conceptualization of Service for Sustainable Coexistence

Helge Löbler

Abstract Service is ubiquitous. It is a fundamental activity in human coexistence. However, many people are unaware that natural and ecological systems provide service as well. For example, 30% of the food consumed by humans depend on natural services, in particular on the pollination by insects. Clean water is preserved by natural water cycles. Human intentions, interests or desires can obviously not explain the existence of all these different kinds of service. They would exist without humans on earth. Consequently, our understanding of service has to be much broader than just including man-made service. This chapter proposes a conceptualization of service as a fundamental process of coexistence particularly the coexistence of humans, and more important, the coexistence of humans together with nature. Today, humans' ecological footprint on earth results in a resource overshoot already in August. On August 8, 2016, we had used as much from nature as our planet can renew in the whole year. We need a better conceptualization of our coexistence on this planet. Humans cannot survive if they destroy the resources they need. However, the planet can survive without humans. Service is proposed and conceptualized here as an integrated concept for a mutual coexistence of humans together with nature. Once we understand the cyclic intertwinement of human and natural activities, humans can adjust their life to natural cycles without reducing quality of life.

Keywords Service • Resources • Exchange-change process • Entropy reduction • Ecology

H. Löbler (✉)
Institute for Service and Relationship Management, Leipzig, Germany
e-mail: loebler@wifa.uni-leipzig.de

S. Barile et al. (eds.), *Social Dynamics in a Systems Perspective*, New Economic Windows, DOI 10.1007/978-3-319-61967-5_6

6.1 Ubiquity of Services

Service is ubiquitous and it has been written much on service and services. It has been studied and defined in various disciplines and can be understood as a general phenomenon in human coexistence. Following the service-dominant logic, service is "the application of skills and knowledge for the benefit of another party" (Vargo and Lusch 2004, 2008). Both, service-dominant logic as well as service science the conceptualization of service is limited to the man-made world: "[Service science] restricts its attention to artificial, in the sense of being real human-made, worlds, and is ran thus a specialization of systems science" (Spohrer 2009, p. 13). Accordingly, service is understood as a means to "value co-creation". Therefore people wants and/or needs explain the phenomenon of service in the human spere. Service is understood as based on needs and desires, as intentional acts and intentional processes made by humans or their technologies (Akaka and Vargo 2014). As far as this relates to the man-made world, there is nothing wrong. However, service is not just a man-made phenomenon, it also exists in the natural world, for example, between organisms and lower or higher developed species. There are other disciplines of service research outside of management, IT, service management and marketing such as biology and ecology, which cover this natural service. For instance, symbiosis research focuses on service not related to humans (Boucher 1985b; Douglas 1994, 2010) and addresses the question of why "different types of organisms help each other" (Boucher 1985b, p. 1). Another field of research where service plays an important role is the field of ecology (Boyd and Banzhaf 2007).

Ecological services are (not intentionally) offered to humans by the ecosystem (Boyd and Banzhaf 2007), for instance in form of food and fresh water. Thirty percent of the food consumed by humans depend on natural services (in particular on the pollination by insects). The services of nature are by far older than the man-made service, as they already existed before humans were on earth. This means, however, that as a general phenomenon (not only limited to humans), service cannot be explained by human motives or intentions.

For an explanation of service in a general way, a superordinate cause or a superordinate concept is needed. While, on the one hand, such a concept has to exceed the previous focus on the human sphere, it should, on the other hand, not marginalize the research that has emerged in the field of human services provided in the various disciplines. Furthermore, this concept has to explain service by other causes than human wishes, motives or intentions, since these explanations do not hold for simple species let alone organisms.

I start to develop such a concept by distinguishing four areas of service between the human and non-human sphere. The following section describes common features of service, as they are found in the biological, ecological and human service literature. Hereafter, I propose a conceptualization of service and a definition respectively which is an extended version of Hill's fundamental definition of service (Hill 1977, 1999). This definition focuses on service as a process of change, however, not every change can be understood as service. If so, the term service would not draw a useful distinction between what service is and what it is not and every change or transformation, positive and negative ones, would be service. Hence, a further specification is needed which is applicable to natural and human service.

For this specification the concept of entropy will be used. The concept of entropy is here used as a measure of disorder and negentropy is used as a measure of order. The core idea of the concept of service, as presented here, is that if service did not exist, disorder would increase according to the second law of thermodynamics. Therefore, service is defined as a process to maintain or increase order against forces of disorder. The reduction of entropy against forces of disorder (for example through maintaining or increasing order) requires open systems in which low entropy states can be achieved by exchanges with the relevant environment. Entities enabeling states of low entropy emerge either in dissipative structures (Prigogine and Nicolis 1967; Prigogine and Leferver 1968) or through autocatalytic hypercycles (Eigen and Schuster 1979) far from equilibria and they are maintained by all living entities as they are open systems (Schrödinger 1944). We usually see human wishes, interests and intentions to be the forces against disorder. However, one can doubt that in the human sphere wishes or interests are intended deliberately if one follows Schopenhauer's statement: "You can do what you want, but you cannot want what you want" (Schopenhauer and Ebeling 1978–1979). Finally, I will describe and calculate entropy reduction through service.

6.2 Four Realms of Services

Although humans are part of nature, I distinguish humans from nature whereby nature is the part which would exist without humans. By doing so, four realms of service can be distinguished: Service exchanged between non-human beings (nature to nature); service provided by nature to humans (e.g. ecosystem service), service exchanged between humans and finally service by humans for nature (see Fig. 6.1).

The first realm of service includes all services transferred by non-humans; this service is provided by nature for nature and often discussed under the term symbiosis (Lewis 1985; Janzen 1985; Boucher 1985a). Different categorical systems have been used to describe different kinds of symbiosis (Starr 1975; Lewis 1985; Connor 1995). Authors agree that in these kinds of interactions, "one of the species provide some kind of 'service' that its partner species cannot provide for itself" (Yamamura et al. 2004, p. 421).

The second realm of service describes all service provided by nature for humans; these are ecosystem services (not to be confused with service ecosystems). Ecosystems provide service such as storm protection and pollination. Pollination of crops by bees is required for 15–30% of U.S. food production; most large-scale farmers import non-native honey bees to provide this service. (Kremen 2005). "Ignoring these services in public and private decision making threatens our ways

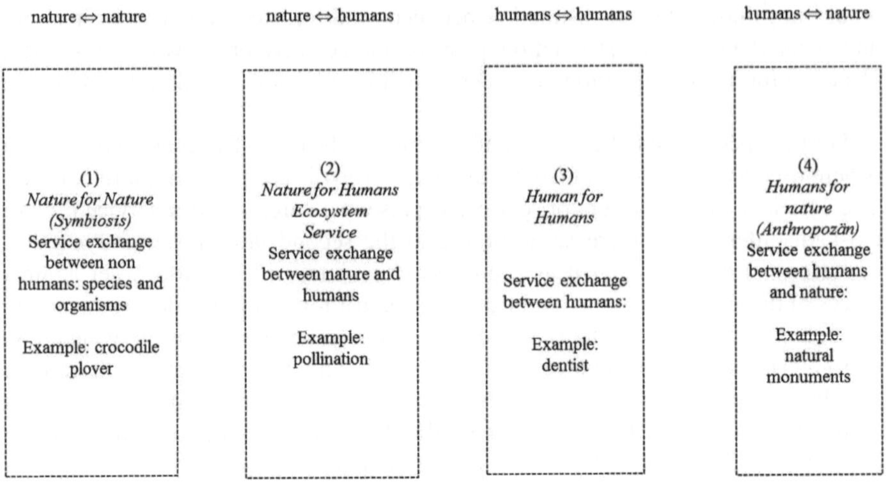

Fig. 6.1 Realms of service

of living and impedes our ability to achieve our aspirations for the future" (Ranganathan et al. 2008, p. 2). Humans benefit from a manifold of resources and processes that are offered by natural ecosystems. While environmentalists have discussed ecosystem services for decades, these services were popularized and their definitions formalized by the United Nations 2004 Millennium Ecosystem Assessment (MEA) (2005), a 4-year study involving more than 1300 scientists worldwide.

The third realm, service exchanged between humans, is not described here, because it is the best known realm.

The fourth realm does not only include preservation of natural heritage, it also covers those natural service, which is now replaced by human's work in specific regions. In Europe, for example, already 40% of the bee colonies have disappeared. In China, there are only 10% left. People there take this threat seriously for man and nature and have started trials for artificial pollination. What are the common denominators of all these different kinds of service?

6.3 Similarities of Human and Natural Services

An extended review of different streams of literature served for identifying four joint denominators for human and non-human service, for instance (Douglas 2010) for symbiosis, (Boyd and Banzhaf 2007) for ecosystems (Vargo and Lusch 2004, 2008, 2011) as well as (Maglio and Spohrer 2008) for human service.

The term service in the field of biology and ecology (Cushman and Whitham 1991; Herre et al. 1999; Yamamura et al. 2004; Ollerton 2006) is firmly established. In these disciplines, the exchange of services is sometimes not distinguished from the exchange of goods: "Mutualisms usually involve the direct exchange of goods and services (e.g. food defense and transport)" (Herre et al. 1999, p. 49). Since service-dominant logic has not yet been integrated into the biology and ecology, this distinction is very similar to the previous distinction between goods and services in the goods-dominant logic (Vargo and Lusch 2004; Vargo et al. 2010).

Using goods-dominant logic language, goods and services in connection with symbiosis can be closer described as follows:

- Services: distribution of seeds, etc., protection, defense, bioluminescence, cleaning etc.
- Goods: carbohydrates, nitrogen, inorganic components, organic components, water etc. (Ollerton 2006)

In service-dominant logic terminology, however, these are only two types of service: the direct and indirect service (Vargo and Lusch 2004, 2008). The goods can thus be understood as indirect service and the "services" as direct service.

6.3.1 Resource Integration

Getting access to resources is one of the most important foundation of the evolution of symbiosis (Kiers and Denison 2008; Kiers et al. 2010). All natural processes are integrating resources without which they would not exist (West et al. 2007; Paszkowski 2006; Noe and Hammerstein 1995).

The notion of resource integration is also a fundamental concern in service-dominant logic as written in FP 9 (Vargo and Lusch 2004, 2008): All (economic and social) actors are resource integrators. There cannot be an activity without resource integration. Resource integration is an ongoing process, "a series of activities performed" (Payne et al. 2008) by an actor and often, but not necessarily, going hand in hand with value co-creation (Ngo and O'Cass 2009; Berghman et al. 2006; Golfetto and Gibbert 2006; Löbler 2013; Peters et al. 2014).

Resources are highly dynamic functional concepts; **they are not, they become,** they evolve out of the triune interaction of nature, man, and culture. Here nature sets outer limits, but man and culture are largely responsible for the portion of physical totality that is made available for human use (Zimmermann 1951, pp. 841–851; Vargo and Lusch 2004).

Whereas Zimmermann refers to physical entities, Vargo et al. extend his view to non-physical entities: "In fact, resources such as time, weather and laws, which are often considered exogenous and uncontrollable by individuals and organizations, are often integrated—if not relied on—in the value creation process by all service systems" (Vargo et al. 2010, p. 148).

Hence, everything in an actor's environment can become a resource depending on the context in which the actor acts. Things become resources, if they are integrated by interaction (Ballantyne and Varey 2006) or other activities. All activities, service in particular, are recourse dependent on resources and can hence only be performed if resources are integrated. This fundamental principle is not limited to the man-made world. Hence, natural processes and human activities are based on resources and their integration and with them all kinds of service. Therefore, the first common denominator is:

Use or Integration of Resources All kinds of service can only be performed by using some kind of resource whether these resources are material (land, seeds, food, etc.) or immaterial (sunlight, information, wind, etc.).

6.3.2 Transfer or Exchange

If natural processes and human activities depend on resources, natural or human actors must be able to have access to these resources. Access to resources (if the actor does not produce it himself) is possible either through an exchange with other actors or through a transfer directly from the environment. The resources obtained from the environment are also produced or provided by other natural or human activities: "[...] mutualisms usually involve the direct exchange of goods and services (e.g. food defense and transport)" (Herre et al. 1999, p. 49). Access to oxygen is consequently an exchange of resources:

It needs no further comprehensive remarks that the concept of exchange, both in economics and marketing, plays a central role. Without exchange, no coexistence is possible. Service-dominant logic focuses on the concept of exchange by notions like "service-for-service exchange" and service "the fundamental unit of exchange" (Vargo and Lusch 2004, 2008; Vargo et al. 2008). Therefore, exchange or transfer forms are another defining characteristic of service.

Transfer/Exchange of Resources Getting access to resources through exchange of resources between the actors and/or with the environment.

6.3.3 Transformation or Change

Transformation or change is another denominator of service. Riddle, in explaining Hill's definition of service (Hill 1977, 1999), identifies service as activities for change: "Service are activities that produce changes in persons or the goods they possess" (Riddle 1986). The integration or use of resources 'produces' changes in the state of the receiver of these resources. This concept of service goes back to Hill:

"The service may be defined as a change in the condition of a person or other goods belonging to the same economic unit, which is brought about is the result of the activity of some other economic unit, with the prior agreement of the formal person or economic unit" (Hill 1977, p. 318).

In the same vein, Sampson (2010) as well as Fromm and Cardoso (2015) define service as a change in the state of the service receiving entity. Although 'change' or 'transformation' is mentioned incidentally in service-dominant logic and service science, there is neither a reference to Hill nor to Riddle let alone a discussion of this concept as being central to the understanding of service. Examples are: Resources are an "ability to cause desired change" (Vargo and Lusch 2008, p. 7), or: "Service systems are value-creation networks composed of people, technology, and organizations. Interventions taken **to transform** state and coproduce value constitute services" (Maglio et al. 2006, p. 81). However, this concept goes back to Hill and is not taken into account by service-dominant logic. Service-dominant logic defines service "as the application of skills and knowledge for the benefit of another party". That the benefit may be caused by a change in the state of the service recipient is not mentioned or discussed in service-dominant logic.

So far the condensed components from social sciences' literature, marketing and IT in particular for a service definition are exchange and change. Thereby, exchange is neither limited to goods, activities or rights, nor is change limited to persons or things. It may be helpful to think of different kinds of service in day-to-day life:

- Haircut: The hairdresser exchanges the application of his/her skills and knowledge for the client's money and changes the appearance of the client.
- Transportation (a bus ride): buying transportation, exchange of a right to use transportation for money, and the transportation changes the place of people.
- Software: Buying the software is exchanging a right (and perhaps some material object) for money and working with the software changes the process of a user's work.
- Car: Buying a car means exchanging a car for money, it changes the owner's situation as she can now driving the car, talking about the car and looking at the car. Here we see how the indirect service (masked by goods) works.
- Renting a flat: Exchanging the right to use the flat for money; change of life conditions.
- Consulting: Exchanging the right of using information or getting information for money; changing the way of thinking or deciding.

In symbiosis research, Douglas (1994, 2010) has summarized the *outcomes* gained by service and Ollerton (2006) has described the *services exchanged* in symbiotic relationships. The detailed description of these resource integrations (Douglas 2010, p. 12; Ollerton 2006, p. 413) show that they are all used for changes and/or transformations in/of organisms or other entities like fixation, respiration, degradation, production, protection, removal etc. (Douglas 2010, p. 14).

Transformations are realized by integrating resources received from another party. Resources are used to transform the state of the receiver compared to the state without integrating or using those resources. Value may or may not emerge if a transformation is realized depending on the context. A third mutual characteristic of man-made and non-man-made service therefore is that resources are integrated for a specific transformation or change. Whether the transformation is beneficial for an entity is determined by the context and not by the resources or by the transformation. Transformation or change is always defined in relation to a situation without

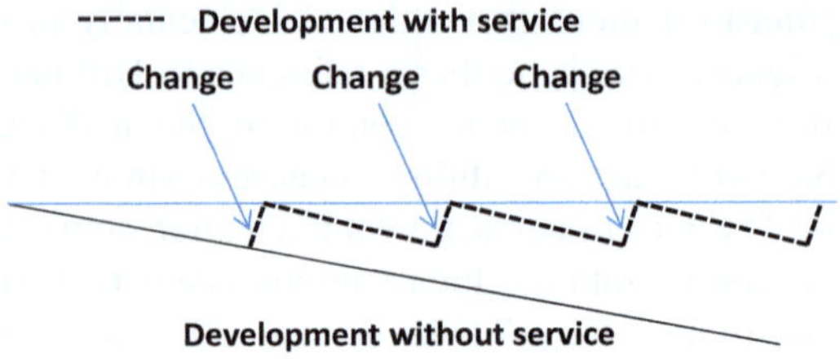

Fig. 6.2 Relative notion of change

the service in question. Hence, if, for example, a service is a maintenance service, maintenance is a transformation/change compared to the situation without the maintenance service.

A second important point has to be mentioned: One might think that innovation has no place in this frame, however, innovation can be simply conceptualized as second order change that is a change of change or a transformation of a transformation. When typewriting was invented, it was a change of the way in which ideas were transformed into letters (Löbler and Lusch 2014). Hence, innovation is a second order change.

One has to be aware that change is never absolute, but always relative. If a condition of a service receiver is getting worse without a service, than a service keeping the condition constant would be a relative change compared to letting the condition get worse. It is usually done in small steps as indicated in Fig. 6.2.

In general, whenever one wants to identify or measure change, one needs a dimension of description or a scale which remains unchanged.

Transformation (Change) of the Receiver's State by the Use of Resources Resources are not integrated (used, consumed) for their own sake but for a transformation/change in a service receiver's state, whereby the receiver usually (but not always) also transforms (consumes or wears down) the resources.

6.3.4 Context Dependence of Value and Survival

In a human world (co-)creation of value is the ultimate reason or goal of service (Vargo and Lusch 2011) and service science is understood as "the study of value co-creation" (Spohrer and Maglio 2010, p. 158).

Simultaneously, it is argued that the application of skills and knowledge can only offer value propositions but cannot deliver value itself (Vargo and Lusch 2008, p. 8): "Enterprises can offer their applied resources for value creation and collaboratively (interactively) create value following acceptance of value propositions, but cannot create and/or deliver value independently" (ibid, p.7). Service providers may intend to deliver value but value can neither be delivered nor can it be determined by the service provider. "Value is always uniquely and phenomenologically determined by the beneficiary" (ibid, p.7.). Value emerges out of the use in a specific context (Chandler and Vargo 2011). Hence, value is contextual and not a defining trait of service, which is rather the application of skills and knowledge. Although value appears as the ultimate reasoning for service, and although the potential of concepts such as value in use (Vargo and Lusch 2004, 2008, 2011) or value as a function of human experience (Ramaswamy 2011) are recognized as important, they are still in their "research infancy" (Ostrom et al. 2010, p. 26). The notion of value-in-context recently introduced by (Vargo and Lusch 2008, 2011) and Chandler and Vargo (2011) emphasizes the context dependency of value. By doing so, value is not understood as inherent or defining characteristic of service. Service may be a necessary condition for value creation, but it is not sufficient to co-create value. This is supported by an emerging interest in the literature on value (co-)destruction (King and Burgess 2008; Plé and Chumpitaz 2010; Echeverri and Skalén 2011) pointing out that value does not necessarily go hand in hand with service. "Following Plé and Chumpitaz (2010), it suggests that service relationships or exchanges, as defined in S-D logic, do not necessarily result in value co-creation, but that value co-destruction may occur too." (Lefebvre and Plé 2011) Hence, value co-creation is not defining element of service as it can also turn out as value "co-destruction". As a consequence of its contextuality, value and value co-creation are not an inherent characteristic of a man-made service. It can emerge with service, but not necessarily. Consequently, value and value co-creation can now, in a first step, be excluded from a service definition of the man-made service. As it is now demonstrated, the same applies for a definition of non-man-made service; the service in nature.

Research on symbiosis analyzing the benefits and harms of symbiotic relationships is aware of the contextuality of these benefits and harms: A "fundamental problem is the variability of real associations, such that benefit is not a fixed trait of some relations but varies with environmental circumstance" (Douglas 2010, p. 6). Organisms which are usually harmless or beneficial can be deleterious to their partners, depending on the context they live in (Douglas 2010). Examples are the fungus Colletotrichum magna which was first identified as a virulent pathogen of certain plant species. However, its impact on plant growth was found to depend on plant species and even cultivar (Redman et al. 2001). In the same vein the Helicobacter pylori, a bacterium in the human stomach, can cause ulcers and gastric cancer especially in older people, but can be beneficial in children providing protection against diarrhea and asthma (Blaser and Atherton 2004). Thus benefit and value are contextual. In the biological as well as in service-dominant logic's understanding of service benefit and/or value are/is dependent of the kind of use and the context of use.

Contextuality of Value (Benefit) of Service Value or benefit is not inherently a service characteristic. Value can emerge via resource integration depending on the relationship between a service receiver and its environment. Hence, value as well as survival is contextual.

Since value is firstly context dependent in the human sphere and secondly an anthropocentric attribution to nature made by humans, it is not taken as a defining characteristic of service as understood here.

So far, the three common characteristics of natural and human service can be condensed into the following sentences: The main function of service is a change of state of a specific entity emerging through resource integration, where these resources are acquired by exchange. Thereby, not only the state of the specified entity is changed, but also the resources which are used in the process of change. Humans transform natural resources into resources which are used by other humans or which are transformed into waste. Fortunately, nature often transforms our waste back into resources, but unfortunately not always. There is an ongoing process of exchange of resources between humans and between humans and nature. It is important to understand the ties between those entities that perform the activities. I use the term 'entities' because not only people are executing activities (transformations and changes) but also animals and plants and also the physical part of nature, for example wind and rain. The ongoing process of transfer or exchange is complemented by a process which happens between the exchanges or transfers. It is complemented by transformation or change.

However, not all transfer-transformation or exchange-change processes are service. If so, the term service would not draw a useful distinction between what service is and what it is not. In this case, everything or every process would be service. Hence, a further specification is needed which is applicable to natural and human service. This will be discussed in the next section.

6.4 Specific Exchange-Change Processes as Service

For practical purposes, I call those processes in which an exchange (transfer) and a transformation of resources, which changes the state of an entity, takes place, an exchange-change processes (ECP). As mentioned above, not all ECP can be defined as service. If all ECPs were defined as service, no processes would be excluded by the definition. A definition that includes everything would lose meaning. Therefore, only certain ECPs are understood as services, and these processes have an additional characteristic. Such additional characteristics can be discovered in various ECPs.

Human and nature are similarly integrated in such ongoing ECPs. Many of them are so common that we are often not aware of them. For instance, the carbon dioxide-oxygen cycle is an ongoing process in which plants provide resources for humans and animals, while humans and animals provide resources for plants in return. CO_2 (carbon dioxide), H_2O (water) and energy (sunlight) are converted during the photosynthesis in O_2 (oxygen) and $C_6H_{12}O_6$ (glucose) (see Fig. 6.3). Oxygen and glucose are the most important resources for humans and animals. Humans and animals use this resource and convert it into carbon dioxide and energy (work) in return. Other ongoing transfer-transformation processes are, for example, the carbon, nitrogen and water cycle. This clearly shows that without transformations or changes between the transfers, a world would have quickly exchanged all resources.

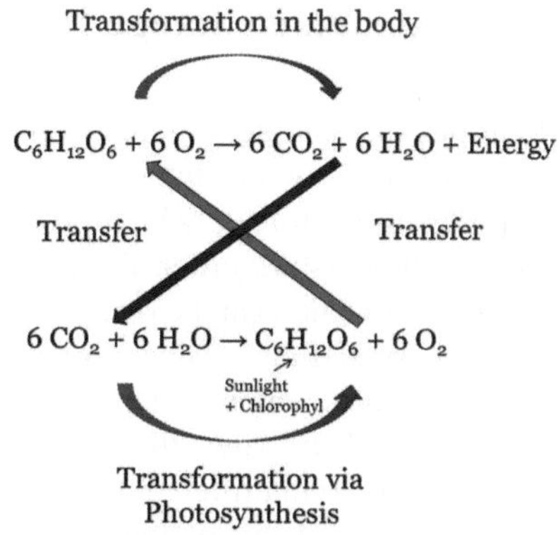

Fig. 6.3 Exchange-change cycle

The transformations in these ongoing ECPs are life-sustaining. The entities, for which the transformations are taking place, use the resources to maintain their inner structure. They are open to integrate resources and they are closed with respect to the structure they maintain. If they were closed with respect to resources, they would not be able to maintain their inner structure. To maintain their inner structure means to also defend the inner order against destroying forces. To do so, they use and need resources they obtain from other entities or from their environment. In short, the resources received from others or from the environment are used to maintain an ordered state. Hence, all transformations and changes can now be distinguished according to whether they maintain or increase order, or whether they reduce it.

Those EPCs, that maintain or increase order, are defined as service. Service thus means to act against disorder or decay in well-defined entities through a transfer and a transformation of resources which is an ECP. So not all ECPs are service; only those which oppose disorder or decay. Service are those ECPs, which counteract disorder or decay in a well-defined entity.

A general measure of disorder caused by a process in an entity is entropy. Entropy can be used as an instrument for the measurement of information (Shannon 1949/1998). Furthermore, entropy can be a measurement of that part of energy, which cannot be converted back—the part, which is irreversible. An intuitive interpretation of entropy is, to use it as an indicator of disorder (Sethna 2011): According to the second law of thermodynamics, closed systems strive for a state of maximum entropy. In an open system, however, entropy can be reduced by increasing the entropy of the environment through the integration of resources.

Following this logic, service is now understood as an entropy reducing transformation or change. Entropy as a measurement can thus be used to characterize the service transformation in detail. With this in mind, services can also be defined as an exchange-change processes that counteract the entropy production in certain entities. One can say that for such entities, the entropy reducing change (transformation) is a service.

6.5 Exchange-Change Processes and (Dis)order

If entities like people, organisms or other creatures would be left without access to resources, they would not be able to counteract decay or disorder. For practical purposes, those forces, which promote decay or disorder, are called disturbances, because they disturb order. For example, the incidence of hunger feelings is such a disturbance, which can be overcome by integrating resources in form of food. Also fatigue is such a disturbance that can be overcome by sleep.

In addition, compared to a desired (but not yet reached) state, the actual state appears as a disturbance that has to be overcome. If the old car is not good enough anymore, it is interpreted here as a disturbance in relation to a new car. A disturbance is always defined in relation to a reference state, which can either be the status quo or a desired state. The desired state can only be reached by the integration of resources. Figure 6.4 shows how resource integration works against disturbances. Disturbances are indicated by D, the resources by R, the change of the state of an entity by C ('C Dis.' indicats a change producing disorder and 'C Ord.' indicats a change producing order) and the states before service and after service are indicated by S_{BS} and S_{AS} respectively. The process shown in Fig. 6.4 starts with a state after a service that is disturbed. Through the disturbance a service is needed; a state which is indicated in Fig. 6.4 as 'before service''. Such a service, or in other words, a resource integrating change, then produces a state after a service again. Whereas the disturbance rises the disorder, the service rises order, which is indicated by 'Dis.' and 'Ord.'. The change of the state is accompanied by a change (transformation) of resources. These resources might become either resources or disturbances depending on the entity receiving them.

Resources are therefore necessary to deal with certain disturbances. The term 'state' is used in a very general sense. It does not only cover static situations, but also routine processes operated by the entity. The disturbed status can therefore be understood both as a static state, but also as a dynamic process. For example, if the process of digestion is disturbed, which can manifest itself in stomach pain, then one would like to see a doctor for his/her service. In this case, the disturbance refers to a process. The process is the status quo to be restored by the service. Very often

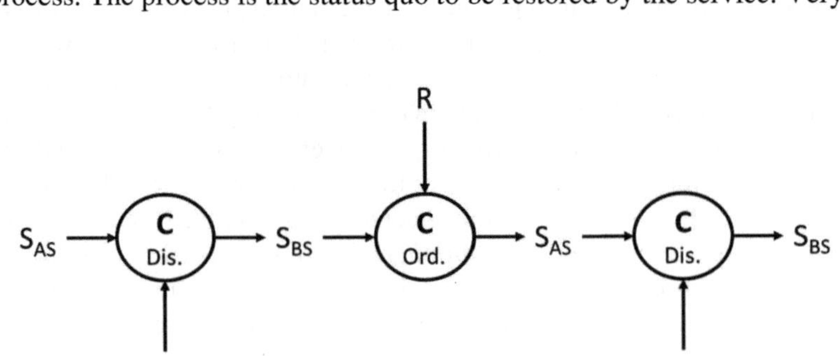

Fig. 6.4 One part of an ongoing process

disturbances relate to processes, such as the precipitation of machines that perform certain operations (processes). The machine is a resource that transforms other resources (materials) and one wants to maintain the machines ability to perform this process. Hence, as long as the machine works, there is no need to change/transform the machine. However, in the case of a machine failure, the machine has to be transformed to work appropriately. In one case, the service is done by the machine, in the other case, it is done upon the machine. Generally, the change of resources and the change of the status of the resource integrating entity go hand in hand. The result is either a changed resource, a changed state of the entity or both.

It is important to understand that Fig. 6.4 represents only a small part of an ongoing ECP. Therefore, in Fig. 6.5 the ongoing process is indicated by two parts.

Figure 6.5 shows that the result of a change in the ongoing ECP can lead to either a new resource or a disturbance depending on the context and entities in question. If, for example, a music-band plays for a garden party, the music is a service for those enjoying the party. Simultaneously, the same music can be a disturbance for the neighbor. Depending on how the neighbor wants to overcome the disturbance, he/she needs specific resources to transform his/her state. The indication of D and R in Figs. 6.4 and 6.5 do not only indicate one resource or one disturbance, but a set of resources or disturbances. The simplicity of the picture should not obscure the fact that humans, other animals or parts of nature, are involved in a variety of ECPs that have been described as service-dominant networks (Löbler 2013). Without change between the exchanges, the exchanges were completed quickly. When resources would only be exchanged but not changed, they would sooner or later all be exchanged and there is no need for further exchanges. This holds for both the static level of the resource exchange as well for the dynamic. Dynamic resources are

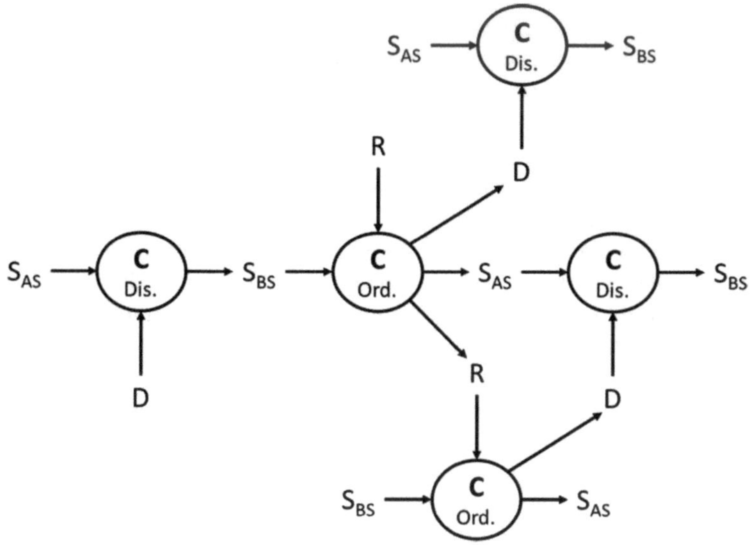

Fig. 6.5 The intertwined ongoing process

processes, which in turn are used as resources. If these processes are changed, it is either a maintenance repair or an innovation. In the first case, the process was disturbed and the status quo as to be restored. In the second case, the improvement of the process serves as a benchmark for an innovation and as a reference state. Compared to this innovation, the old process is a disturbance. In any case, an innovation is the change of a change; or a second order change.

In nature, innovations correspond to mutations, which in turn means a change in the process controlling genes. For all these transformations in an entity, both in an object and in a process, the entropy can be calculated.

6.6 Transformation as a Reduction of Disorder/Entropy

6.6.1 Without Use of Benefit or Value Units

To illustrate the entropy changes of the changed states a change matrix serves as a starting point (see Table 6.1 (a) for human service and (b) for natural service).

On the left hand side there are different disturbances which can be overcome by using the resources on the top of the matrix.

There is no limitation of what disturbances, resources and transformations can be. Table 6.1a shows well known disturbances and options to overcome these disturbances. Table 6.1b shows disturbances that are overcome by nature. In Table 6.1 the results of the transformation are indicated by an 'a' for acceptable and an 'n' for unacceptable. However, each cell of the transformation matrix can contain any kind of result. Each individual cell within the matrix may be further differentiated. So, for example, hunger can be satisfied in very different ways and with very different food. Therefore, resources and outcomes vary accordingly.

Later, we will use concrete numbers indicating values or benefits. But firstly, the entropy of a transformation is calculated without using numbers to indicate value or benefit.

The entities considered here, as shown in Table 6.1 are principally open in respect to entropy reducing resources. Disturbances, as understood here, are all forces that increase the entropy, if the entity does not counteract by using resources.

How does the entropy of a state change, if induced by a disturbance or by a service? If an entity is in a state without any disturbance, this state is denoted by the term 'reference state'. This state can be challenged by disturbances. If this reference state is challenged by one disturbance, then the state of the entity is described by the reference state and the disturbance. The situation is now described by the reference state and one disturbance, a set of two elements. The more disturbances are challenging the reference state, the more elements the set has that describes the situation. In general, the state of an entity is defined as a set $S = \{s_0, s_1, s_2, s_k, .., s_K\}$ where s_0 refers to the reference state an K to the number of disturbances. The cardinality of this set is $|S| = K + 1 = M$.

Table 6.1 Change-Matrix [(a) Example humans and (b) Example nature]

(a) Change-Matrix (Example humans)

			R (resources)		
	C (change)	Hairdresser	Food	Beverage	Medical Doctor
	Hunger	n	a	n	n
D (disturbances)	Old look compared to a desired new one	a	n	n	n
	Thirst	n	n	a	n
	Qualmishness	n	n	n	a

(b) Change-Matrix (Example nature)

			R (resources/ processes included)		
	C (change)	Global hydrological cycle	Photosynthesis	Activity of microorganisms in the soil	Aposematism[a]
	Polluted water	a	N	n	n
D (disturbances)	Too much CO_2 in the air	n	A	n	n
	(Compostable) waste	n	N	a	n
	To become eaten (animals)	n	N	n	a

[a]Aposematism is a widely used function of bioluminescence, providing a warning that the creature concerned is unpalatable

Now this situation is described by the entropy measure by using relative frequencies of the elements of the describing set to calculate entropy of S (Shannon 1949/1998). The entropy of a state described by M unweighted elements is $H(S) = -\sum_1^M \frac{1}{M} \operatorname{ld}\left(\frac{1}{M}\right)$.[1] In this case $H(S) = \operatorname{ld}(M)$.

Table 6.2 shows the results of entropy rise induced by more and more disturbances.

Reading Table 6.2 from the bottom to the top shows the entropy reduction if disturbances are eliminated or overcome.

In this way, each contribution of a transformation that eliminates or overcomes a disturbance can be described as entropy reduction.

The procedure indicated in the example shows, that the entity's state becomes more ordered the more disturbances are eliminated or overcome by means of service.

Accordingly, entropy can be understood as a measure of the adequacy of a service. Since such entropy calculations are not restricted, it can be applied to all transformations and thus represents a general measure of the adequacy of service as order-creating change of an entity's state. This is an understanding of service being independent of constructs derived from the human sphere like value, benefits or alike. In nature, we cannot identify human sphere categories like value or benefit. However, with entropy we propose a general measure of the adequacy of service. Simultaneously, the calculation of entropy can include human notions as value and benefits, as shown in the next section.

Table 6.2 Entropy of different states

Number of disturbances	Cardinality of the describing set	Formula	Entropy
0	1	ld(1)	0
1	2	ld(2)	1
2	3	ld(3)	1585
3	4	ld(4)	2
4	5	ld(5)	2322

[1] ld(x) denotes $\log_2(x)$, the logarithm to the base 2. To reflect a higher order sensitivity, smaller bases may be chosen, because it is about order and not about information, in which the base 2 assured the unit "bit". When measuring disorder, the basis indicates how strong an increase in the disorder is perceived, when the number of options increases by 1. The base 2 indicates that a doubling of the possibilities increase the "disorder unit" of one. A base of for example 1.5 indicates then one and a half time the options would be perceived as a disorder increase of one "disorder-unit". Since the perceived disorder increases with decreasing gains, the base is always greater than one. One can speak of a diminishing marginal disorder in relation to the number of options. 1 divided by the basis is a measure of the increasing sensitivity to perceive opportunities as disorder. This is evident because a reduction of order from two options to one is relatively much more than a reduction from e.g. ten to nine.

6.6.2 *With Use of Benefit or Value Units*

As in Sect. 6.6.1 the state of an entity is described by a set of elements which are the reference state and the disturbances $S = \{s_0, s_1, s_2, \ldots s_k, .., s_K\}$. Furthermore, it is assumed that the reference state as well as the elimination or overcoming of the disturbances is evaluated by the entity. Since the disturbances are defined in relation to the reference state, it is reasonable to assume that the reference state is associated with the highest value. The reference state is either a desired state or the status quo, so that its value is always higher, if compared to less desired states. The disturbances are assumed to be independent from each other.

In correspondence with the set of describing elements $S = \{s_0, s_1, s_2, \ldots s_k, \ldots, s_K\}$ the values are

$$W = \{w_0, w_1, \ldots w_k, \ldots w_K\} \text{ with } w_1, \ldots w_k, \ldots w_K < w_0$$

For calculating entropy, we define the following weights reflecting the values: $u_k = \frac{w_k}{\sum_0^K w_k}$.

According to Straathof (2007), the weights u_k can be used like relative frequencies for entropy calculation.

The entropy of a state then is calculated as $H(S) = -\sum_0^K u_k \, ld(u_k)$.

If, for example, an entity is faced with two disturbances and evaluates the reference state and the disturbances with $w_0 = 500$; $w_1 = 350$ and $w_2 = 75$, the weights u_k are $u_0 = 0.541$, $u_1 = 0.378$ and $u_3 = 0.081$ and $H(S) = -0.541 \, ld(0.541) - 0.378 \, ld(0.378) - 0.081 \, ld(0.081) = 1.304$.

Compared to the situation without values, where the entropy is 1.585, the entropy with values is lower, because the values reveal a preference structure indicating a higher level of order than without a preference structure. If the values for the three elements are equal, e.g. 300, they would not reveal a preference structure and would hence indicate a lower level of order with an entropy of 1.585.

Furthermore, if now the entity can overcome one of these disturbances, the values matter. If the disturbance with the value of 350 can be eliminated, the entropy is $H(S) = 0.559$, whereas if the disturbance with the value of 75 is eliminated, the entropy is $H(S) = 0.977$. If the disturbance valued with 350 remains in the set, the values are 500 and 350, indicating a less clear distinction of the elements compared to set with the values of 500 and 75. The letter case shows a clearer distinction in the revealed preferences and therefore a higher level of order indicated by a lower entropy.

In this way, all possible constellations can be calculated and described using entropy: Both situations in which no benefit or value units are assigned, as well as those cases, in which values or benefits are given. In general, service can be construed as entropy-reducing transformations of resources to overcome disturbances, referring either to a status quo or to an intended state. The more entropy a service reduces, the more appropriate it is.

6.7 Conclusion

This article assumes that service is provided not only in the sphere of human activity. Rather, it takes the view that service is ubiquitous even without humans, as it is discussed in biology and ecology. Therefore, this article intends to understand and to conceptualize service as a general process, which can be explained without reference to constructs of the human sphere like interests, whishes or values. However, this does not mean that those human constructs are excluded in the conceptualization proposed here.

For the theoretical underpinnings of the process of service, it was purposive to locate service between exchanges of resources and to conceptualize it in the vein of Hill's perspective as a change either of the state of the resource receiving entity or the resources themselves. Changes in general can now either rise or reduce entropy. As in closed systems entropy automatically increases, only those transformations/changes are understood as service that reduce entropy in a well-defined entity. Therefore, service is not explained by intentions, desires, or the like, but as energy transformations in open systems which are typically accompanied by matter transformations to protect the entity "from decay" (Schrödinger 1944). This protection, however, is in the non-human sphere a result without human intention. Whether a transformation protects an entity in nature, can be seen only ex-post and not ex-ante. In nature, entropy reducing entities are created in dissipative structures (Prigogine and Nicolis 1967; Prigogine and Leferver 1968) or in autocatalytic hypercycles (Eigen and Schuster 1979) far from equilibrium. Service thus can be understood as a fundamental process of maintaining order (in the broadest sense). Therefore, service is of such fundamental importance, because it counteracts the forces that augment the disorder. It was demonstrated by examples how transformations counteract entropy rising forces and how entropy can be used as a measure of disorder in order to assess such transformations/changes.

Consequently, entropy can be understood and used as a useful concept to assess the adequacy of services; regardless of whether the results are assed by humnas' value or benefit or not.

Furthermore, the broader concept of service as understood here, is a basis for understanding the cycles of humans' intertwinement with natural processes. We, as humans, may become smarter, but we are still a part of nature and we are not independent from natural laws. If humans accept and understand the fundamental laws of coexistence, they can live on earth without damaging and/or exploiting nature. Service as a resource integrating, cyclic activity, as proposed here, is the common ground for coexistence on earth.

References

Akaka, M. A., & Vargo, S. L. (2014). Technology as an operant resource in service (eco)systems. *Information Systems and e-Business Management, 12*(3), 367–384.

Ballantyne, D., & Varey, R. J. (2006). Creating value-in-use through marketing interaction: The exchange logic of relating, communicating and knowing. *Marketing Theory, 6*(3), 335–348.

Berghman, L., Matthyssens, P., & Vandenbempt, K. (2006). Building competences for new customer value creation: an exploratory study: Creating value for the customer through competence-based marketing. *Industrial Marketing Management, 35*(8), 961–973.

Blaser, M. J., & Atherton, J. C. (2004). Helicobacter pylori persistence: Biology and disease. *Journal of Clinical Investigation, 113*(3), 321–333.

Boucher, D. H. (Ed.). (1985a). *The biology of mutualism – Ecology and evolution*. New York: Oxford University Press.

Boucher, D. H. (1985b). The idea of mutualism. In D. H. Boucher (Ed.), *The biology of mutualism – Ecology and evolution* (pp. 1–28). New York: Oxford University Press.

Boyd, J., & Banzhaf, S. (2007). What are ecosystem services? The need for standardized environmental accounting units. *Ecological Economics, 63*(2–3), 616–626.

Chandler, J. D., & Vargo, S. L. (2011). Contextualization and value-in-context. How context frames exchange. *Marketing Theory, 11*(1), 35–49.

Connor, R. C. (1995). The benefits of mutualism: A conceptual framework. *Biological Reviews, 70* (3), 427–457.

Cushman, J. H., & Whitham, T. G. (1991). Competition mediating the dynamics of a mutualism: Protective services of ants as a limiting resource for membracids. *The American Naturalist, 138*, 851–865.

Douglas, A. E. (1994). *Symbiotic interactions*. Oxford: Oxford Science Publications, Oxford University Press.

Douglas, A. E. (2010). *The symbiotic habit*. Princeton, NJ: Princeton University Press.

Echeverri, P., & Skalén, P. (2011). Co-creation and co-destruction: A practice-theory based study of interactive value formation. *Marketing Theory, 11*(3), 351–373.

Eigen, M., & Schuster, P. (1979). *The hypercycle: A principle of natural self-organization*. Berlin: Springer.

Fromm, H., & Cardoso, J. (2015). Foundations. In J. Cardoso, H. Fromm, S. Nickel, G. Satzger, R. Studer, & C. Weinhardt (Eds.), *Fundamentals of service systems, service science: Research and innovations in the service economy* (pp. 1–32). Cham: Springer International Publishing.

Golfetto, F., & Gibbert, M. (2006). Marketing competencies and the sources of customer value in business markets: Creating value for the customer through competence-based marketing. *Industrial Marketing Management, 35*(8), 904–912.

Herre, E., Knowlton, N., Mueller, U., & Rehner, S. (1999). The evolution of mutualisms: Exploring the paths between conflict and cooperation. *Trends in Ecology and Evolution, 14* (2), 49–53.

Hill, T. P. (1977). On goods and services. *Review of Income and Wealth*, (4), 315–338.

Hill, T. P. (1999). Tangibles, intangibles and services: A new taxonomy for the classification of output. *The Canadian Journal of Economics*, (2), 426–446.

Janzen, D. H. (1985). The natural history of mutualism. In D. H. Boucher (Ed.), *The biology of mutualism – Ecology and evolution* (pp. 40–99). New York: Oxford University Press.

Kiers, E. T., & Denison, R. F. (2008). Sanctions, cooperation, and the stability of plant-rhizosphere mutualisms. *Annual Review of Ecology, Evolution, and Systematics, 39*(1), 215–236.

Kiers, E. T., Palmer, T. M., Ives, A. R., Bruno, J. F., & Bronstein, J. L. (2010). Mutualisms in a changing world: An evolutionary perspective. *Ecology Letters, 13*(12), 1459–1474.

King, S. F., & Burgess, T. F. (2008). Understanding success and failure in customer relationship management. *Industrial Marketing Management, 37*, 421–431.

Kremen, C. (2005). Managing ecosystem services: What do we need to know about their ecology? *Ecology Letters, 8*(5), 468–479.

Lefebvre, I., & Plé, L. (2011). *Emergence of value co-destruction in B2B context*. Paper presented at the Naples Forum on Service 2011, Vol. 2011.
Lewis, D. H. (1985). Symbiosis and Mutualism: Crisp concepts and soggy semantics. In D. H. Boucher (Ed.), *The biology of mutualism – Ecology and evolution* (pp. 20–39). New York: Oxford University Press.
Löbler, H. (2013). Service-dominant networks. *Journal of Service Management, 24*(4), 420–434.
Löbler, H., & Lusch, R. F. (2014). Signs and practices as resources in IT-related service innovation. *Service Science, 6*(3), 190–205.
Maglio, P. P., & Spohrer, J. C. (2008). Fundamentals of Service Science. *Journal of the Academy of Marketing Science, 36*, 18–20.
Maglio, P. P., Srinivasan, S., Kreulen, J. T., & Spohrer, J. (2006). Service systems, service science, SSME, and innovation. *Communications for the ACM, 49*(7), 81–85.
Ngo, L. V., & O'Cass, A. (2009). Creating value offerings via operant resource-based capabilities. *Industrial Marketing Management, 38*, 45–59.
Noe, R., & Hammerstein, P. (1995). Biological markets. *Trends in Ecology and Evolution, 10*, 336–339.
Ollerton, J. (2006). Biological Barter: Patterns of specialization compared across different mutualists. In N. M. Waser & J. Ollerton (Eds.), *Plant-pollinator interactions – From specialization to generalization* (pp. 411–435). Chicago: The University of Chicago Press.
Ostrom, A. L., Bitner, M. J., Brown, S. W., Burkhard, K. A., Goul, M., Smith-Daniels, V., & Rabinovich, E. (2010). Moving forward and making a difference: Research priorities for the science of service. *Journal of Service Research, 13*(1), 4–36.
Paszkowski, U. (2006). A journey through signaling in arbuscular mycorrhizal symbioses. *New Phytologist, 172*(1), 35–46.
Payne, A. F., Storbacka, K., & Frow, P. (2008). Managing the co-creation of value. *Journal of the Academy of Marketing Science, 36*(1), 83–96.
Peters, L. D., Lobler, H., Brodie, R. J., Breidbach, C. F., Hollebeek, L. D., Smith, S. D., Sorhammar, D., & Varey, R. J. (2014). Theorizing about resource integration through service-dominant logic. *Marketing Theory, 14*(3), 249–268.
Plé, L., & Chumpitaz, R. (2010). Not always co-creation: Introducing interactional co-destruction of value in service-dominant logic. *Journal of Services Marketing, 24*(6), 430–437.
Prigogine, I., & Leferver, R. (1968). Symmetry breaking instabilities in Dissipative Systems. II. *The Journal of Chemical Physics, 48*(4), 1695–1700.
Prigogine, I., & Nicolis, G. (1967). On symmetry-breaking instabilities in dissipative systems. *The Journal of Chemical Physics, 46*(9), 3542–3550.
Ramaswamy, V. (2011). It's about human experiences. . .and beyond, to co-creation. *Industrial Marketing Management, 40*(2), 195–196. (Special issue on service-dominant logic in business markets).
Ranganathan, J., Raudsepp-Hearne, C., Lucas, N., Irwin, F., Zurek, M., Karen Bennett, K., Ash, N., & West, P. (2008). *Ecosystem services – A guide for decision makers*. World Resources Institute.
Redman, R. S., Dunigan, D. D., & Rodriguez, R. J. (2001). Fungal symbiosis: From mutualism to parasitism, who controls the outcome, host or invader? *New Phytologist, 15*, 705–716.
Riddle, D. L. (1986). *Service-led growth. The role of the service sector in world development*. New York: Praeger Publishers.
Sampson, S. E. (2010). The unified service theory. A paradigm for service science. In P. P. Maglio, C. A. Kieliszewski, & J. C. Spohrer (Eds.), *Handbook of service science, service science: Research and innovations in the service economy*. Boston, MA: Springer US.
Schopenhauer, A., & Ebeling, H. (1978–1979). Die beiden Grundprobleme der Ethik: Behandelt in 2 akad. Preisschr, Philosophische Bibliothek, Bd. 305–306, Meiner, Hamburg.
Schrödinger, E. (1944). *What is life*. Cambridge: Cambridge University Press.
Sethna, J. P. (2011). *Entropy, order parameters, and complexity*. Oxford: Clarendon Press.

Shannon, C. E. (1949/1998). The mathematical theory of communication. In C. E. Shannon & W. Weaver (Eds.), *The mathematical theory of communication* (pp. 29–125). Urbana, IL: University of Illinois Press.

Spohrer, J. (2009). Service science and systems science: Perspectives on social value. Institute of Technology, Tokyo. In *Proceedings of the 5th Symposium of the 21st Century COE Program "Creation of Agent-Based Social Systems Sciences"* (pp. 9–33).

Spohrer, J., & Maglio, P. (2010). Toward a science of service systems – Value and symbols. In P. P. Maglio, Ch. A. Kieliszewski, & J. C. Spohrer (Eds.), *Handbook of service science* (pp. 157–194). New York: Springer.

Starr, M. P. (1975). A generalized scheme for classifying organismic Associations. *Symposia of the Society for Experimental Biology, 29*, 1–20.

Straathof, S. M. (2007). Shannon's entropy as an index of product variety. *Economics Letters, 94* (2), 297–303.

Vargo, S. L., & Lusch, R. F. (2004). Evolving to a new dominant logic for marketing. *Journal of Marketing, 68*(1), 1–17.

Vargo, S. L., & Lusch, R. F. (2008). Service-domiant logic: Continuing the evolution. *Journal of the Academy of Marketing Science, 36*(2), 1–10.

Vargo, S. L., & Lusch, R. F. (2011). It's all Be2B...And beyond: Towards a systems perspective of the market. *Industrial Marketing Management, 40*(2), 181–187.

Vargo, S.L., Lusch, R.F., & Akaka, M.A. (2010). Advancing service science with service dominant logic – Clarifications and conceptual development. In P. P. Maglio, C. A. Kieliszewski, & J. C. Spohrer (Eds.), *Handbook of service science* (pp. 133–156). New York, NY: Springer.

Vargo, S. L., Maglio, P. P., & Akaka, M. A. (2008). On value and value co-creation: A service systems and service logic perspective. *European Management Journal, 26*(3), 145–152.

West, S. A., Griffin, A. S., & Gardner, A. (2007). Evolutionary explanations for cooperation. *Current Biology, 17*, R661–R672.

Yamamura, N., Higashi, M., Behera, N., & Wakano, J. Y. (2004). Evolution of mutualism through spatial effects. *Journal of Theoretical Biology, 226*, 421–428.

Zimmermann, E. (1951). *World resources and industries* (2nd ed.). New York, NY: Harper & Row.

Chapter 7
In Anticipation of Black Swans

Raul Espejo

Abstract Problematic situations are recurrent in organisational systems. We are all, individually or collectively, managing them and need strategies for this purpose. Indeed, for as long as we maintain interactions with others we are managing problematic situations; sometimes we simplify them too much yet in others we respond chaotically without proper reflexion; in either case the cost of these responses to people and organisations can be too high. In this paper I reflect upon a third strategy aided by Ashby's Law of Requisite Variety and complexity theory. The law gives insights to manage environmental and organisational complexities and complexity theory gives insights about environmental constraints and the emergence of self-organisation. Together they provide a view of organisational systems and most significantly, of how to improve the management of extreme situations - *black swans*-, such as wars, catastrophes, social unrest, extremism, and multiple other expressions of very high variety situations.

Keywords Requisite variety • Complexity theory • Constraint • Black swans • Self-organisation

7.1 Pareto's 80/20 Rule

In Heart of Enterprise Beer (Beer 1979, p. 15) gives the example of the reorganisation of the railways in England in the 1960s. He uses Pareto's 80/20 rule (see Fig. 7.1). A revision of the railways profitability roughly confirmed that 20% of the tracks were responsible for 80% of the pay-offs and, management assumed that by focusing on this most profitable 20% the railways performance could significantly improve. Proposals were made for closing down the inefficient tracks. However, this restructuring did not anticipate that closing the 80% less efficient tracks only implied shifting the burden to a new tail of unprofitable tracks (see Fig. 7.2) and taken this to its extreme they had found a strategy to reduce the railways to its

---

R. Espejo (✉)
Syncho Research, Flat 3, 30 Nettleham Road, Lincoln LN 2 1RE, UK
e-mail: r.espejo@syncho.org

S. Barile et al. (eds.), *Social Dynamics in a Systems Perspective*, New Economic Windows, DOI 10.1007/978-3-319-61967-5_7

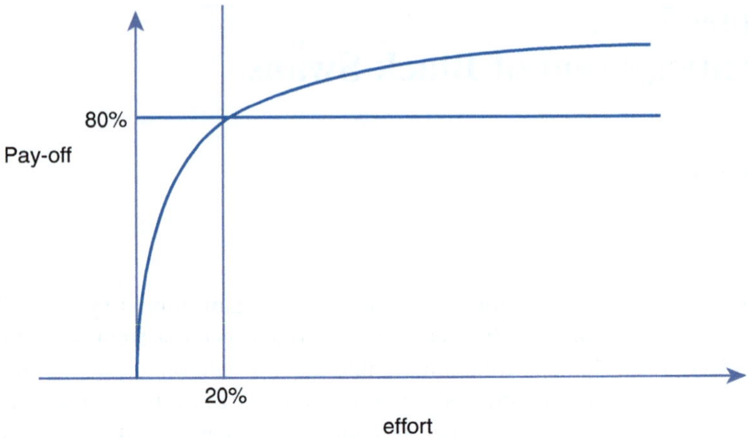

Fig. 7.1 A Pareto curve of effort vs. pay off

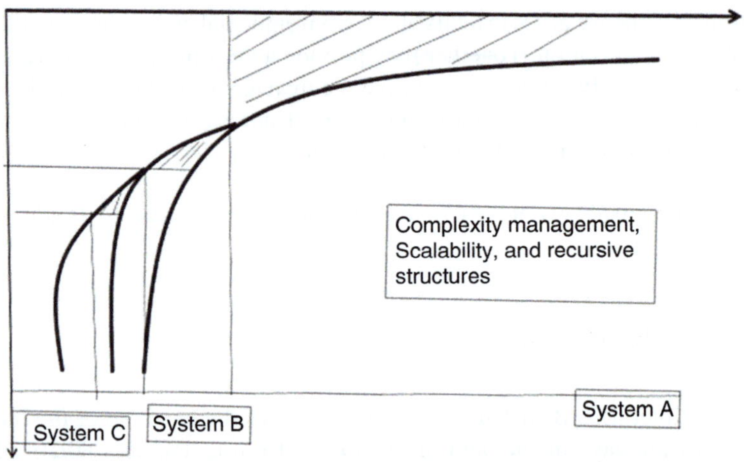

Fig. 7.2 A machine for eating the railways. Elaboration from: S. Beer (1979, p. 17)

minimum expression. Pareto curves are common in all kinds of situations and their conceptual underpinnings are becoming increasingly clear with power laws (PL).

Increased efforts for smaller pay-offs are common. We observe this convexity in economic activities, in people's and organisations' learning curves and more generally, in system-environment interactions. However, contrary to the idea of a "machine for eating the railways" as implied by the English railways example, Fig. 7.2 could be understood as "a machine for creativity and innovation" to cope with high complexity events. As efforts approach diminishing returns in an increasingly complex environment the pressure to amplify the system's own response capacity increases. The fact that events make apparent performance limitation in the current situation may trigger the need for innovation, that is, for fresh new responses. Environmental complexity is already triggering the need for new systemic capabilities (Teece 2008). The argument is that as environmental complexity increases with unexpected events, the appropriate response is, as implied by Ashby's Law of Requisite Variety (Ashby 1964) increasing the variety of the system. However, this proposition triggers a whole range of methodological questions, particularly, does it always imply increasing the variety of the system? But, one way or the other, Pareto's rule anticipates that in an increasingly interconnected world organisational systems will be buffeted by significant, often unknown, events and that unless these systems can cope with unexpected complexity with imagination and creatively their performance will suffer.

7.2 Revisiting the Law of Requisite Variety

Ashby (1964) offered the concept of variety as a measure of complexity; he defined variety as the number of *possible* states of a situation, and the Law of Requisite Variety (LRV) as a regulatory requirement for balanced interactions; "only variety absorbs variety". Usually this is understood as 'the variety of the regulator has to be as large as the variety of the reguland', or the 'variety of the system has to be as large as the variety of the environment'. However, this matching of varieties is more subtle than it is implied by these statements. Figure 7.3 illustrates the interactions between a system and its relevant environment (Espejo and Howard 1982). For a system to maintain stability within an acceptable *level of performance* or *target set* (V_T) the *disturbances or variety of environmental events* (V_D) must be matched by the system's affordances or *response variety* (V_R).[1] This performance is regulated by a regulator. Regulation in general is distributed throughout the system, as the regulator may be constituted by many self-regulating regulators (only for clarity Fig. 7.3 shows one external regulator to the system). In addition to V_D, V_R and V_T we need to consider the *variety of possible behaviours or outcomes* (V_O) of the system to anticipate control capacity. Is V_O within the set of acceptable behaviours V_T?

The LRV tells us first, that the ratio V_D/V_R is larger or at best equal to the variety of outcomes (V_O) and second, that for the regulator to have requisite variety, this variety must be contained by the *target set variety* (V_T). For human activities this target set (V_T) relates to the regulator's purposes and values. But Fig. 7.3 also helps appreciating that to achieve requisite variety the regulator needs *learning schema or models* to change controllable variables towards reducing errors between outcomes and targets. But good models and behavioural schema of the interactions between a system's actors and relevant environmental agents are not enough to change

[1] I understand affordances as *resource* supported interactions between a system's actors and environmental agents, which allow the system producing requisite responses to environmental disturbances.

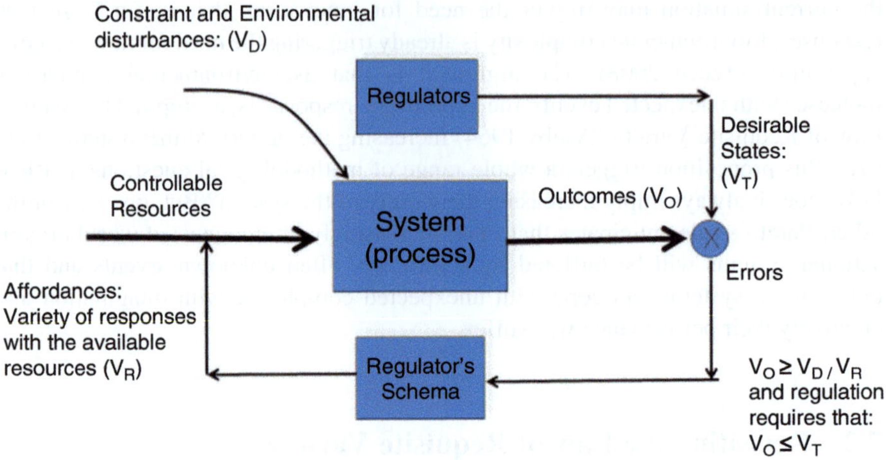

Fig. 7.3 Scalability, innovation and structural recursion

controllable variables; additionally, the regulator needs response capacity—affordances—to produce this learning. Resources are necessary to produce requisite responses. As an extension of the LRV, Conant and Ashby (1970) proved the theorem: "Every good regulator of a system must be a model of that system". Just it is necessary to keep in mind that without resources the regulator lacks response capacity.

Schema may help the regulator to work out systemic and environmental constraints but does not give requisite response capacity.

Environmental constraints, rather than chaotic, unconstraint possibilities, is what makes learning worth in a situation. In his book *An Introduction to Cybernetics* Ashby (1964, p. 134) says: "... learning is worthwhile only when the environment shows constraints". Furthermore in p. 247 he says "the variety in the disturbances V_D is not really as large as it seems... the disturbances show a constraint." Thus the case is the following: V_D has many components, each of which shows variety. The first estimate of V_D's variety puts it very high, and we are in danger of deducing (if the regulator's capacity is given) that regulation to a required degree of performance is not possible. Further examination of V_D may, however, show that the components are not independent, that constraints exist, and that the real variety in V_D is much lower than the first estimate. It may be found that, with V_R's capacity given, this smaller variety can be regulated, and full regulation or control achieved..." Thus the *discovery and implementation of a constraint* may convert "regulation impossible" to "regulation possible". If V_R's capacity is fixed, it is the only way.

On the one hand if V_T is smaller than V_O, possible outcomes are beyond the acceptable states and therefore sooner or later the situation will be out of control. No doubt the stringent is the performance criterion the less likely is that the regulator will possess control capacity and it can be anticipated that control will fail sooner or later. On the other hand, from the above arguments, a regulator can visualise and make use of environmental constraints to transform "regulation impossible" into "regulation possible". Schema plays an important role in achieving this regulatory capacity. For instance, for a company, climate regulations can be seen as environmental regulatory constraints to reduce the chances of destructive weather disturbances in its operations.

For high variety, environmental events (V_D), climate change requires high variety responses (V_R) in order to maintain people's quality of life (V_T). Unusual changes may imply high variety outcomes (V_O) i.e. floodings, and if the system's affordance or response capacity are inadequate the ratio V_D/V_R will be high compared with the variety of desirable states making the variety of possible outcomes (V_O) large and the system's regulators will lack requisite variety. V_D/V_R will be high, and the LRV tells us that the variety of possible outcomes (V_O) will exceed the variety V_T of acceptable states. In this case the system will not have requisite variety to respond to extreme disturbances. Therefore, unusual but possible events, like extreme flooding, will pose high risk to the system.

7.3 Cybernetic Explanation, Constraint, and Co-evolution

From an epistemological perspective, cybernetic explanations are negative: "Causal explanation is usually positive. We say that billiard ball B moved in such and such direction because billiard ball A hit it at such and such angle. In contrast to this, cybernetic explanation is always negative. We consider what alternative possibilities could conceivably have occurred and then ask why many of the alternatives were not followed, so that the particular event was one of those few which could, in fact, occur." (Bateson 1973, p. 375). Cybernetic explanation is focused on the constraints that discard possibilities and limits the requirements for regulation. The Law of Requisite Variety offers a powerful heuristic to visualize the need for constraints. *A system can adapt just so far as its environment is constrained, and no further* (Ashby 1964, p. 127). However hard a regulator works the outcomes of any situation lacking in requisite variety will sooner or later be out of control. Indeed, if the situation remains unchanged, an observer can anticipate that it will hit difficulties and in the end will be unsuccessful. Comparing the regulator's response capacity with the variety of possible disturbances allows a trained observer to say "there is no way that in the longer run outcomes will remain within the target set". *Thus, working out environmental constraints and designing systems to benefit from those constraints has profound social implications.* The challenge is matching their varieties at acceptable levels of performance. In a more directive sense environmental regulation is also a way of constraining its variety; regulated markets constrain disturbances for economic agents.

System's design is often necessary to match environmental variety with requisite response capacity. If performing a system's task, say its policies, does not recognise in its structure environmental constraints it may proliferate unnecessary variety, making the system's regulation unmanageable; on the other hand, chunking this

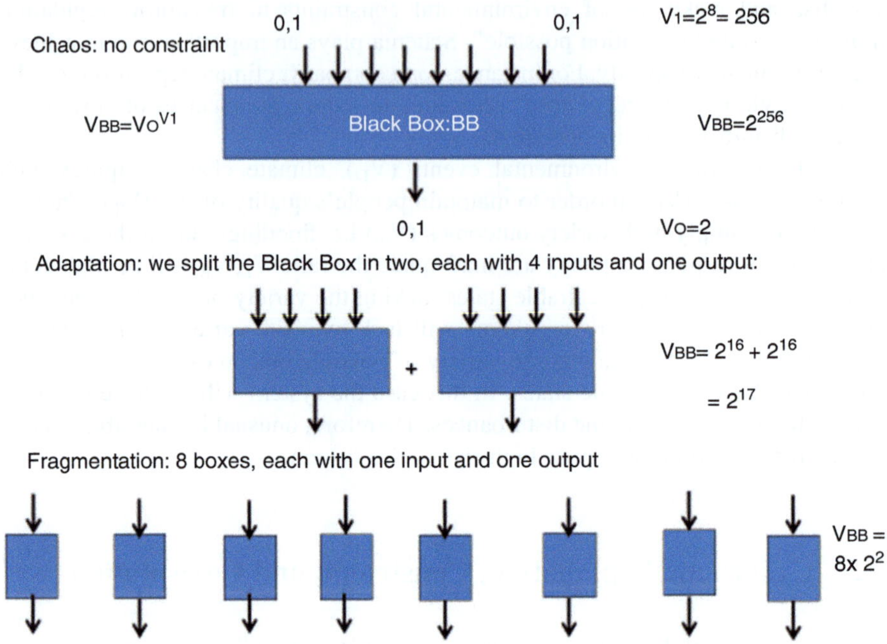

Fig. 7.4 Strategies to manage complex situations. Sources: S. Beer (1979, pp. 47–48)

task by mirroring environmental constrains, the system's complexity can be reduced in orders of magnitude (see Fig. 7.4).

Figure 7.4 illustrates important complexity management strategies for a black box in its environment. Interactions proliferate at unimaginable levels even in simple situations. For instance, with reference to Fig. 7.4, a black box with eight inputs, each with two possible states, 0 or 1, and 1 output also of two possible states, can generate 2^{256} possible states over time, which is an astronomic number of possible states (Beer 1979). Unconstrained variety proliferation makes tasks unmanageable. However, if the black box is divided into two black boxes each with four inputs and one output, the overall variety is reduced to 2^{17} (Fig. 7.4) making the situation more manageable. This variety can be further reduced if the black box is fragmented into eight black boxes each with one input and one output. Of the above three complexity management strategies, the first one, with a structure of unconstraint connectivity is chaotic. Regulators of the black box most likely will lack response variety for the system to achieve a desirable performance in its environment. The third strategy, of eight boxes, oversimplifies the situation making unlikely adequate responses to environmental disturbances; it offers a fragmented schema. However, the second strategy, of two boxes, may map better environmental constraint and offer a more adaptive strategy. The first strategy, the unconstrained *holistic* strategy, may be chaotic and beyond control. The third, the *fragmented* schema, is inadequate and may not permit the necessary connectivity to perform a task well. The second, intermediate strategy is more interesting; it is an indication of possible adaptation. The two black boxes are indeed complex themselves but they may offer more hope if they succeed mapping environmental constraints. Assuming that the division in two is mapping relevant constraints in the environment, thus concentrating response capacity where the pressure is, the variety absorption of these two boxes will deal with most of the environmental variety leaving a small residual variety to the attention of those responsible for the integration and coordination of the overall task (of the box with eight inputs and one output); this strategy can give requisite variety to the regulator "a system can adapt just so far as its environment is constrained, and no further". We are now moving in the direction of making manageable a so far chaotic situation. Regulation of the total situation—the larger box—will focus attention on the cohesion of the two autonomous black boxes to achieve a performance that makes possible the global task.

Chunking a task requires learning and adaptation to overcome fragmentation. However, for as long as the system mirrors relevant, perhaps so far hazy, environmental constraints, distributing response capacity among autonomous but interconnected chunks is indeed a good strategy to make a situation more resilient. For a system, a large complexity can be an asset but only if it is underpinned by structural constraints that mirror environmental constraints. The less we understand and exploit these constraints the more difficult will be to produce desirable tasks and achieve high performance. *Creativity can be interpreted as a discovery of constraints.* Based on these ideas is that it is possible to distinguish between senseless fragmentation and purposeful complexity unfolding (Espejo and Reyes 2011). As said before, to have a good regulator it is necessary (but not sufficient) to be a model of the regulated situation (Conant and Ashby 1970). In other words, for an organisational system to perform well in its relevant environment it must be a model of that environment. To achieve a desirable task the organisational system needs to work out constraints to achieve requisite variety. This is a challenge to creativity and innovation. The embodiment of the environmental constraints must be in the unfolding of the system's complexity, that is, in the splitting of the black box complexity into chunks of complexity that increase adaptability, resilience and reduce the complexity of the situation to manageable levels. Indeed, fragmentation happens when this break down is done without a good grasp of constraint in the environment.

Often environmental constraints are not recognised and if they are recognised this happens after the event as an outcome of painful errors and not of anticipatory creativity. In other words, often systems handle poorly the structural mirroring of relevant environmental complexity. To illustrate this point we need go no further than the economic and financial systems in recent times. Swings between centralisation and decentralisation have dominated the design of complexity management strategies. The tacit strategy of centralisation is that "bosses" know better. On the other hand decentralisation assumes that people will find their way better without regulatory interference. In few words we can say that the first strategy aims at attenuating social complexity through hierarchical structures, while the second aims at a proliferating social complexity by reducing controls and assuming that self-regulation and self-organisation will make the trick. Unfortunately, both strategies have produced social and economic fragmentation, the first mostly by imposing unnecessary restrictions and the second by weak regulations. Complex social systems are networks often underpinned by incompetent fragmentation or by fostering dangerous connectivity. For instance, the economic system requires articulation and co-evolution with financial services in its environment. Financial services are environmental enablers of the economy, and following Conant and Ashby, to achieve economic development at an acceptable level of performance, mutual co-development and co-regulation must follow economic policies and map the economy's unfolding of complexity from the local to the global; economic aspects must intertwine with financial aspects at all structural levels. The strategy of having structurally large financial services with emphasis on the global economy, that is, financial services dominated by large international banks weakly coupled to local economic agents can be seen as responsible for the economy's weak unfolding of its complexity. The environmental constraints experienced by production enterprises do not strengthen their co-evolution with financial services, which are more interested in their own viability. We can expect that this structural arrangement does not help a healthy economic development. The large interconnectivity of financial services more focused on their own viability than on the viability of production enterprises increases the chances of a weak complexity unfolding of the economy. Decentralisation of financial services mirroring the economy's commercial enterprises should be beneficial to the global economy in the longer run. As said before a system can adapt just as far as its environment is constrained, and no further.

To summarise, an economic system in an environment of financial enterprises more focused on their own viability than on success of the enterprises constituting the economy, and in need to perform well in competitive markets, are likely to trigger a dysfunctional break down of the economy's tasks, that is, of a dysfunctional unfolding of complexity that increases the chances of less focused enterprises with less stable relationships. Regulators of these relationships need to constrain environmental complexity to give requisite variety to enterprises. The evolution of these financial enterprises as autonomous, dynamic non-linear systems makes them sensitive to small changes and to self-organised criticality, which increase the chances of dangerous cascading effects as it was the case in 2008 (Haldane and May 2011). This case may threaten not only their viability but the viability of the economy as well. In other words, a small addition of risk may produce big unexpected changes as the system reaches its self-organised criticality. With reference to the 2008 financial crisis "a single sub-prime grain produced the self-organised criticality of the financial sector" (Haldane and Nelson 2012), and challenged the stability of the whole economy. Without building appropriate 'walls' (i.e. constraints) crises may spread rapidly throughout the system.

7.4 Power Laws and Organisational Scalability

The strategies of an organisational system to maintain dynamic stability with its environmental agents requires attention to constraints and co-evolution. *A system can adapt just so far as its environment is constrained, and no further* (Ashby 1964), *and the system requires creativity to co-evolve and recognise environmental constraints to increase its adaptability*. This co-evolution was illustrated for the economy's interactions with financial services. Two aspects need attention; on the one hand learning how organisational systems absorb and develop environmental complexity through their structures and behaviours, and on the other hand how environmental agents constrain the variety of their disturbances to the system by self-regulation and self-organisation and also, in policy terms, by guided self-organisation (Gerschenson 2015).

These are exceedingly complex situations dominated by proliferating interactions that cannot be accounted effectively by Gaussian normal distributions; to develop a view of the system's behaviour through aggregations of independent facts is only relevant for linear interactions. Systems interactions with interconnected environmental agents are accounted by often infrequent and high complexity events. The patterns of these events are captured by power laws that follow 'long tailed' Pareto distributions as in Fig. 7.1. The 80/20 rule is a fingerprint of their systemicity. The performance of the English railways depended in more than cutting off the tails. These tails suggest systemic interactions and tacit complexity absorption strategies that can be described by power laws. These laws describe a wide variety of situations. For instance nature absorbs complexity following the "Square/Cube Law" responsible for producing fractals; "In an organism surfaces absorbing energy grow by the square but the organism grows by the cube, resulting in an imbalance; fractals emerge to bring surface/volume back into balance" (Boisot and McKelvey 2013, p. 68). Another power law in natural and also social systems is the form in which organisms and organisations evolve as their components recognise the limits of "*connecting costs*", which trigger the formation of organs and cells in organisms (McKelvey et al. 2012) and units with organisational closure in organisational systems (Beer 1979; Maturana and Varela 1992; Wene 2015). Complexity unfolding, the cascading of autonomous systems within autonomous systems in organisations (Espejo and Reyes 2011) is the outcome of self-organising processes when an autonomous system hit the limits of *connecting costs* and need larger embedding or smaller embedded autonomous systems to deal with the proliferating complexity. Top-down situations require unfolding their complexity into autonomous systems within autonomous systems simply because the global system cannot cope with the unmanageable complexity of interactions with environmental agents. In bottom-up situations, as environmental complexity grows locally, an autonomous local unit may need to coordinate its activities with others to produce a larger organisational system. These are processes that underpin the emergence of very large organisations. *Connecting cost* is accounted by power laws. We know that most enterprises are small and a few reach very large sizes and we anticipate that their distribution follows a power law. Another power law is "self-organised criticality", this happens in situations where non-linearity and amplifications of small changes produce dramatic non-linear outcomes; these are instances of *accidents frozen in time* (Boisot and McKelvey 2013). Woods are instances of situations containing accidents frozen in time; a small fire in a forest may be contained or otherwise depending on the existing fire walls or constraints built up into the forest, suggesting that structural aspects underpin the way the fire unfolds, either to containment or to a natural catastrophe, that is, a high variety event. Similarly, the almost melt-down of financial institutions in 2008 was a case of sub-prime financial products, whose failure could not be contained by the financial system because the banking systems was too interconnected; failure to understand how to deal with the connecting costs of an overblown financial system blinded policy makers to see that they had to build up constraint and support co-evolution with production enterprises. The financial system lacked structural constraints to restrict the damage that small institutions originated in larger financial institutions (Haldane and May 2011). With the support of power laws and organisational cybernetics policy-makers and regulators could have reflected upon the co-evolution of the economic and financial systems and discussed their structural recursion to build up walls to contain the failure of small institutions avoiding reaching the point of banks too big to fail. Good cybernetics of the financial system would have reduced the damage (Espejo 2015). And, with the support of power laws it would have been possible to work out systemic events and environmental behaviours highlighting unconstrained interactions favouring the spreading of the crisis; outliers emerge from social systems and environments in co-evolution.

Figure 7.5 offers an alternative way of describing 80/20 distributions with the use of log-log diagrams (McKelvey and Boiset 2009). It distinguishes the Gaussian and Pareto worlds. Cognitively, people account for small events of high frequency (the left side of the figure) using mean values and standard deviations; this is a way for them to give order to data, on the other hand large and infrequent events (like those in the right side) cannot be accounted unless cognitively we accept systemicity and find empirical evidence fitting power laws; these outliers of different size fit the negative slope of the diagram.

In terms of risk management, we are aware of the uncertainties stemming from the non-linearity of situations dominated by 'butterfly effects'. Social situations are dynamic non-linear systems in which, as illustrated for the finance system, small changes in some of its components may trigger unexpected and large effects in time. These systems are dominated by uncertainty and not by measurable, centrality driven, risk; they may experience unexpected black swans or outliers or extreme behaviours (Taleb 2008). Since it is not possible to anticipate how or when small changes will produce black swans, fitting power laws to already experienced outliers may help anticipating necessary system's capabilities to deal with the unexpected as and when they happen. This anticipation requires organisational systems with response capacity to deal with distributed outliers that fit power law distributions, avoiding the chaotic overloading of a poorly structured organisational system experiencing *connecting cost.* Structural recursion becomes a complexity

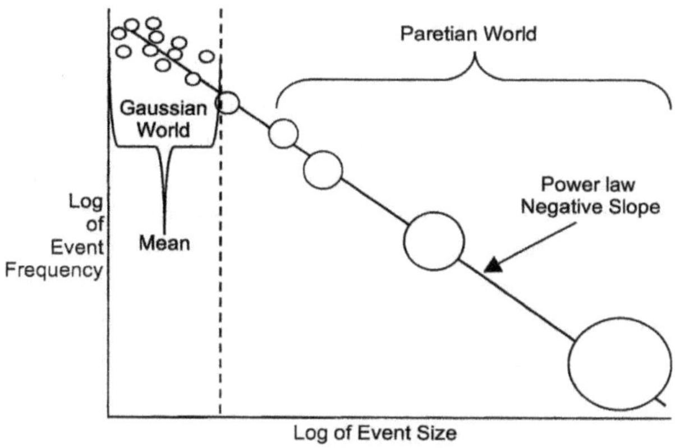

Fig. 7.5 Gaussian and Pareto Worlds. Source: McKelvey (2013) Stylized Pareto distribution on log-log scales

management strategy closely connected to frozen seeds of extreme events. In this perspective, the increasing connectivity of systemic components and their related proliferating environmental complexity may trigger adaptive organisation structures better prepared to deal with the unexpected. *Good cybernetics, supported by empirical research, is a must for policy processes aimed at co-evolutionary processes to improve the response capacity of systems, making them more adaptive and resilient* (Espejo 2015).

What is apparent is that for a system the 80/20 distribution expresses interconnectivity and complexity while the cognitive schema of people working with Gaussian distributions, of averages and standard deviations, imply orderly events which can be managed as aggregations of independent events. However helpful this latter approach might be to deal with already structured situations, it offers an unrealistic view of a complex world. Power laws give us the chance to build up response capacity to unexpected, problematic situations, which reflect systemicity at several levels. These are risky and unpredictable problematic situations with increasing need for creativity and innovation, beyond the standard responses of linearity. They produce Pareto tails that require distributed adaptation and learning. The English railways' attempt to cut off the tail of diminishing returns was a recipe to destroy its complexity and functionality. Pareto distributions of unexpected behaviours recognise connecting costs, self-organised criticality and other forms of systemic behaviour that require high variety responses. These behaviours may indicate high variety events lacking appropriate adaptive responses and possibly systems operating in dysfunctional chaotic regimes or fragmented regimes (the first and third complexity management strategies discussed before in this paper) that fail achieving adequate performance. These would be organisational systems failing to co-evolve and adapt to complex environments. Systems would benefit from policies building up constrains in their co-evolution with environmental agents. For *complex adaptive systems,* these are the hallmark of scalable structures, such as Beer's recursive structures (Beer 1979).

7.5 Scalability

Even small social systems experience *connecting costs*, which, to avoid chaos, require a cascading of autonomous systems. Responses to high variety stimuli require the unfolding of a structure with capacity to respond adaptively to environmental situations. This implies attention to environmental constraints and building up affordances supported by innovative schema and resources. Figure 7.3 offered a view of the interdependence of response requirements and environmental disturbances for effective regulation. Improved and innovative schema helps seeing constraint in chaotic disturbances; it helps attenuating environmental variety and exploring for more affordability (i.e. resources) to achieve requisite variety. No change in schema implies business as usual and makes more difficult finding seeds (chaotic attractors) to structure complexity. Finding out constraints is essential to work out necessary affordances for adequate performance. However, it is not enough to find appropriate schema, it is also necessary to build up capabilities to respond to disturbances. In Fig. 7.3 V_R is the structural and organisational resources *producing* responses to disturbances. It embodies the capacity to produce responses. Without this capacity schema is in the system's informational and not in its operational domain (Espejo 2000). Faced to unmanageable *connecting costs* the system depends on scalability for adaptation and requisite variety. As already said structural recursion is by and large the outcome of self-organizing processes, which may drive bottom-up scalability, like in the case of small communities growing into cities, but also, may drive top-down scalability like in enterprises decentralising their resources.

The Pareto curves of Fig. 7.6 can be visualised as a heuristic to support the scalability of recursive processes. Rather than curves highlighting decreasing returns they can be seen as heuristics to see the need for an organisation's distributed creativity and innovation. Empirically, as decreasing returns take place at one level, growing *connecting costs* make necessary the scalability of resources for the distribution of creative responses to environmental complexity through new schema. This is a heuristic to increase the robustness and resilience of organisational systems, which requires guiding policies of scalability to strengthen processes of self-organisation in the system and its environment.

Dealing with the huge complexity of social situations requires enabling the scalability of organised complexity; capabilities for fluid adaptability and potent cohesive operations within multiple operations at multiple structural levels. This is perhaps, the most powerful strategy to increase response capacity to disturbances and black swans. From the perspective of guided self-organisation we talk about complexity unfolding and adaptation to an environment with constrains; the social situation offers an adaptive emergent complexity at the edge of chaos. Guided self-organisation is the encounter of the top-down and bottom-up emergent complexities. We may consider black swans as crystals in problematic environments that offer opportunities for catalysing self-organisation. Large problematic events may require multiple structural recursions to maintain the situation under control. Black

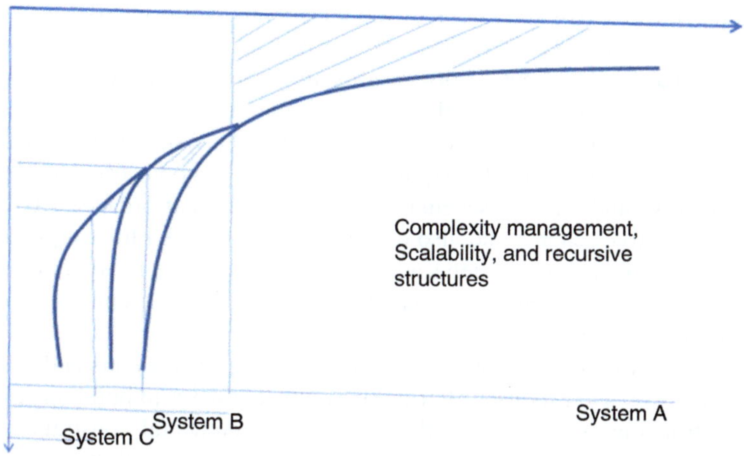

Fig. 7.6 Scalability, innovation and structural recursion. Elaboration from: S. Beer (1979, p. 17)

swans are also distributed events that take place from the global to the local; this is the fingerprint of their structural complexity. Structural scalability reflects the situation. From an empirical position it is important to work out power laws for events potentially disturbing the situation. For example, we may expect that the growth of a city increases potential risks for its citizens and unless the city scales both up and down this growth through recursive structures it will fail providing resilience from the global to the local. The explosion of civil unrest—a black swan affecting a community—may weaken its global fabric. Structural scalability offers constrains to limit the impact of self-organisational criticality.

7.6 Conclusions: An Agenda for Further Research

This paper offers a heuristic for organisational effectiveness; how is it possible for any of our social endeavours to be resilient in highly uncertain environments. The systemicity of our world, the connectedness of all of us in spaceship Earth, makes individual and/or collective social enterprise risky. Constructing effective organisational systems requires making them resilient to the unexpected. These systems have to be prepared to deal with unimagined situations. The fingerprint of complexity in this uncertain environment appears to be power laws that provide anticipations of how complexity organises itself. These power laws are manifestations of constraints in the environment. They don't tell us what is going to happen but they connect butterflies in the Amazon to storms in the northern hemisphere. This complexity, often chaotic, triggers unexpected events and also puts pressure in managing situations at the edge of chaos. In this paper I have argued for increased efforts to learn about these power laws; this is an empirical challenge, since their distributions provide hints about complex adaptive systems, in particular of self-organised criticality as expressions of the constraints that organisational systems need to deal with in this interconnected world. It is constraint that makes more manageable the surroundings of organisational systems and in some cases makes them successful for the good of the people. Supported by the Law of Requisite Variety and its derivation that "Every good regulator of a system must be a model of that system" I have explained the requirements to achieve good performance, which entail innovative and creative schema and well-structured affordances. To achieve distributed adaptive schema and also distributed affordances I have argued for the scalability of organisational systems and in particular for complexity unfolding into autonomous systems within autonomous systems (Beer's recursive structures, Espejo and Reyes 2011).

I have hypothesised that power laws apply to unexpected events and that rather than producing crises responses after these events it is wiser to build up scalability in social structures, supported by discovering environmental constraints and by matching organisational structures, knowing that responses with requisite variety will be necessary. As complex problematic events happen their consequences require responses with effective structural recursion as implied in Fig. 7.6. It should be possible to check empirically black swans anticipating black swans and connecting them to appropriate organisational systems operating at the edge of chaos. This closeness is necessary because maximum emergence of new organisational forms will occur with maximum information i.e. minimum predictability, while minimum emergence will occur with minimum information, i.e. maximum predictability, that is, closer to an ordered regime (Gershenson 2015). In the language of this paper, while the latter may benefit from Gaussian distributions to study behaviour, the former may benefit from power law distributions.

Self-organization has been used to describe swarms, flocks, traffic, and many other systems where local interactions lead to a global pattern or behaviour. These are situations where coordination and cohesion produce new patterns or behaviours, which are the ones that anticipate unexpected events and require new schema and organisation structures. Guided self-organization (Prokopenko 2009; Ay et al. 2012; Polani et al. 2013) should help steering self-organizing dynamics towards a desired configuration of an organisational system (Gershenson 2012).

Black swans are events for which often there is no response capacity, at the cost of people and organisations and the challenge is to anticipate building up capacity to make possible this response. If the event makes apparent the need of a global response, but the response schema is not modified in the system the chances are that new episodes without response capacity will emerge. We may assume that the huge complexity of the situation will trigger events that follow power laws distribution. These are the hypotheses that need further research.

References

Ashby, R. (1964). *An introduction to cybernetics*. London: Methuen & Co Ltd.

Ay, N., Der, R., & Prokopenko, M. (2012). Guided self-organization: perception – Action loops of embodied systems. *Theory in Biosciences, 131*(3), 125–127. doi:10.1007/s12064-011-0140-1.

Bateson, G. (1973). *Steps to an ecology of mind*. St Albans: Paladin.

Beer, S. (1979). *The heart of enterprise*. Chichester: Wiley.

Boisot, M., & McKelvey, B. (2013). Extreme outcome, connectivity, and power laws: Towards and econophysics of organization. In J. Child & M. Ihrig (Eds.), *Knowledge, organization and management* (pp. 61–92). Oxford: Oxford University Press.

Conant, R., & Ashby, R. (1970). Every good regulator of a system must be a model of that system. *International Journal of Systems Science, **1***(2), 89–97.

Espejo, R. (2000). Self-construction of desirable social systems. *Kybernetes, 29*(7/8), 949–964.

Espejo, R. (2015). Good social cybernetics is a must in policy processes. *Kybernetes, 44*(6–7), 874–890.

Espejo, R. & Howard, N. (1982). *What is requisite variety? A re-examination of the foundation of Beer's method* (Working Paper Series No 242). The University of Aston Management Centre.

Espejo, R., & Reyes, A. (2011). *Organizational systems: Managing complexity with the viable system model*. Berlin: Springer.

Gershenson, C. (2012). The world as evolving information. In A. Minai, D. Braha, & Y. Bar-Yam (Eds.), *Unifying themes in complex systems* (Vol. 7, pp. 100–115). Berlin: Springer. URL: http://arxiv.org/abs/0704.0304

Gershenson, C. (2015). Requisite variety, autopoiesis, and self-organization. *Kybernetes Special Issue on the Cybernetics of Self-organisation, 44*(6/7), 866–873.

Haldane, A. G., & May, R. M. (2011). Systemic risk in banking ecosystems. *Nature, 469*, 351–355.

Haldane, A. G., & Nelson, B. (2012). Tails of the unexpected. The credit crisis five years on: Unpacking the crisis. University of Edinburgh Business School.

Maturana, H. R., & Varela, F. (1992). *The tree of knowledge*. Boston, MA: Shambhala Publications.

McKelvey, B. (2013). Reflecting on max Boisot's Ashby space; Applied to complexity management. In J. Child & M. Ihrig (Eds.), *Knowledge, organization and management; Building on the work of Max Boisot* (pp. 93–105). Oxford: Oxford University Press.

McKelvey, B., & Boiset, M. (2009, January). Complexity science: A bridge between modernist and postmodernist perspectives on organizations. *The Academy of Management Review*.

McKelvey, B., Lichtenstein, B. B., & Andriani, P. (2012). When organisations and ecosystems interact: toward a law of requisite fractality in firms. *International Journal of Complexity in Leadership and Management, 2*(1/2), 104–136.

Polani, D., Prokopenko, M., & Yaeger, L. S. (2013). Information and self-organization of behavior. *Advances in Complex Systems, 16*(2, 3), 1303001. doi:10.1142/S021952591303001X.

Prokopenko, M. (2009). Guided self-organization. *HFSP Journal, 3*(5), 287–289. doi:10.2976/1.3233933.

Taleb, N. N. (2008). *The black swan: The impact of the highly improbable*. London: Penguin Books.

Teece, J. D. (2008). *Technological know-how, organizational capabilities, and strategic management: Business strategy and enterprise development in competitive environments*. London: World Scientific Publishing Co.

Wene, C. O. (2015). A cybernetic view on learning curves and energy policy. *Kybernetes, 44*(6/7), 852–865.

Chapter 8
Customer Value Co-creation in a Service-Dominant Logic Perspective: Some Steps Toward the Development of a Measurement Scale

Antonio Botti, Mara Grimaldi, and Massimiliano Vesci

Abstract The work aims at proposing a methodological procedure for value co-creation measurement. Despite the increasing popularity of the concept, in fact, extant research on value co-creation lacks the elaboration of an accepted framework for its measurement strictly complying with Service-Dominant logic's assumptions.

Specifically, the study introduces the different steps of the methodological path that should be followed to elaborate a measurement model for value co-creation and categorizes the potential concrete behavioral and cognitive activities performed by consumers during real exchange process.

The methodology adopted is based on the multi-item procedure developed by Churchill (J Market Res 16:64–73, 1979) which perfectly fits the investigation of multidimensional constructs by ensuring an appropriate semantical coverage. In detail, the first two steps of the model [(1) specification of the domain of the construct; (2) items generation] have been performed.

The originality of the contribution lies in the proposal of a systematization of real value co-creation activities in order to measure the construct in all its semantic facets.

From a theoretical point of view, the chapter provides researchers with a preliminary categorization preparing the ground for the construction of a scale for consumers' value co-creation activities. Moreover, the framework encourages managers to adopt new practical measures to enhance user's involvement at each level and to segment customers to consequently enable and maximize value co-creation behaviors.

Keywords Service-dominant logic • Value co-creation • Measurement model • Multi-item measures • Value co-creation activities

---

A. Botti (✉) • M. Grimaldi • M. Vesci
Department of Business Science – Management and Innovation Systems, University of Salerno, Fisciano, Italy
e-mail: abotti@unisa.it

S. Barile et al. (eds.), *Social Dynamics in a Systems Perspective*, New Economic Windows, DOI 10.1007/978-3-319-61967-5_8

8.1 Introduction

Service-dominant logic (S-D logic, Vargo and Lusch 2004, 2008, 2016) replaces the traditional model of exchange based on the centrality of "goods" (Good-dominant logic, G-D logic). The perspective introduces a *service-for-service* view redefining the "old" conceptualization of service, customer-supplier relationship and the notion of value. In their foundational premises (FPs), in fact, Vargo and Lusch (2004, 2008) maintain that services are drivers for applying users' competences in order to generate mutual benefits for all the actors engaged in value exchange, whereas goods are intended merely as vehicles for service provision (FP1).

In Service-dominant logic value is co-created collaboratively by each stakeholder rather than being created by the firm and then distributed to customers, as in G-D logic. In fact, in S-D Logic the customer is always a co-creator of value (FP6): users are seen as active participants (*actor-to-actor*, A2A) and resource integrators in the final determination of value. This statement implies firstly that value creation is interactional and secondly involves a shift from the primacy of *value-in-exchange* toward the primacy of *value-in-use*. Value is not anymore univocally determined by providers but is negotiated through a joint and interactive exchange process between suppliers, users and other co-creators (FP8). The tenth premise expresses that value is unique to a single actor and can only be determined by that actor. So, it can be affirmed that the notion of value co-creation stands out as the emblem of S-D logic, symbolizing the establishment of a new mindset based on the networked nature of resources exchange and of value co-creation process characterized by the active participation of both producers and customers. In line with the centrality of immaterial aspects of consumption in contemporary service era, this concept has gradually received increasing attention in literature fostering the diffusion of a great number of studies ranging from B2B marketing (Lambert and Enz 2012) to consumer behaviour (Chandler and Chen 2015).

However, despite representing a cutting-edge phenomenon, extant research lacks the elaboration of an accepted framework for the measurement of value co-creation that primly conforms to S-D logic's assumptions (McColl-Kennedy et al. 2012; Payne et al. 2008).

Therefore, the aim of the present paper is to advance a measurement model for value co-creation in order to identify, assess and scale its dimensions in a S-D logic perspective.

In particular, current work attempts to identify the different steps of the methodological path that should be followed to properly measure the construct in line with Churchill's (1979) methodology. Moreover, the paper proposes the dimensions that better describe the concrete behavioral and cognitive activities performed by consumers during the value co-creation process in line with the adoption of an S-D logic approach.

This study focuses on consumer's point of view by addressing the call of McColl-Kennedy et al. (2012), Payne et al. (2008) and Arnould et al. (2006) who suggest further research should better investigate the real actions accomplished by customers during value exchanges. In addition, customers have a leading role in the whole service design and delivery being involved in all the phases of delivery (there are stakeholders such as public administration which are inevitably more strictly related to some specific stages) and directly connected to increase in service effectiveness. In fact, relationships with consumers can determine significant improvements in value offerings and encourage loyalty (Johnson and Selnes 2004).

The identification of the dimensions and sub-dimensions of value co-creation behavior allows to better articulate the construct and leads to a preliminary systematization of the concept which prepares the ground for the construction of a measurement scale. In the second place, it also permits to pinpoint the relationships between value co-creation and other constructs which are particularly relevant in the field of marketing research. From a managerial standpoint, customer value co-creation behavior scale can aid managers in selecting and segmenting customers to enable and maximize value co-creation behavior (Yi and Gong 2013).

The work is structured as follows. In the first section (paragraph 2), the multi-stages procedure developed by Churchill (1979) is described from a theoretical point of view and then a re-elaboration for addressing value co-creation measurement is proposed. The following sub-paragraphs illustrate each step of the model with particular focus on the first two and on the last one (specification of the domain; items generations; identification of antecedents and consequences) which have been also empirically performed. Then, in the last sections the findings, the theoretical and managerial implications and suggestions for further research are discussed.

8.2 Methodology: The Procedure for the Development of a Measurement Scale

Value co-creation is a complex and intangible concept whose investigation cannot exclude the evaluation of different kind of dimensions, such as psychological, social, behavioral and economic, which lead users to develop a given attitude and to perform a given behavior (Edvardsson et al. 2011; Bendapudi and Leone 2003). For this reason, from an empirical point of view, value co-creation translates into a variable not directly observable.

According to Churchill (1979), in fact, marketing variables often refer to aspects of people's psychic life and cannot be measured by simply asking users to express a clear judgement on the phenomenon. These variables, defined latent variables, measure "constructs" that are ideas developed to allow the categorization and description of some directly observable behaviours (Crocker and Algina 1986). The latent variables are inferred from a set of observable indicators so it is necessary to identify measures of constructs that are directly observable.

Therefore, the methodology introduced by Churchill (1979), designed to inspect immaterial dimensions, perfectly complies with the measurement of a multidimensional construct such as value co-creation. In particular, the use of multi- item procedure allows to take into account the complexity of this variable through the development of a multi-phase process which guarantees its appropriate semantical coverage and which ensures and increases scale validity and reliability step by step.

Particularly, in his seminal work the author suggests six stages (eight stages with the two phase of data collection) for developing good measures for multi-item measures[1]: specify domain of the construct, generate sample of items, (collect data), purify the measure, (collect data), assess reliability, assess validity, develop norms.

Over the course of time, a series of reelaboration of the model—which organize the procedure in sub-stages or macro-stages—have been proposed (Bagozzi 1980; Bearden et al. 1993; Smith et al. 1996). Starting from previous contributions, as Fig. 8.1 shows, the framework herein developed is composed of five stages, subdivided in turn into three categories.

The first macro-area (*semantical construction and assessment*) is devoted to the theoretical definition of the construct which contemplates, in the first place, a

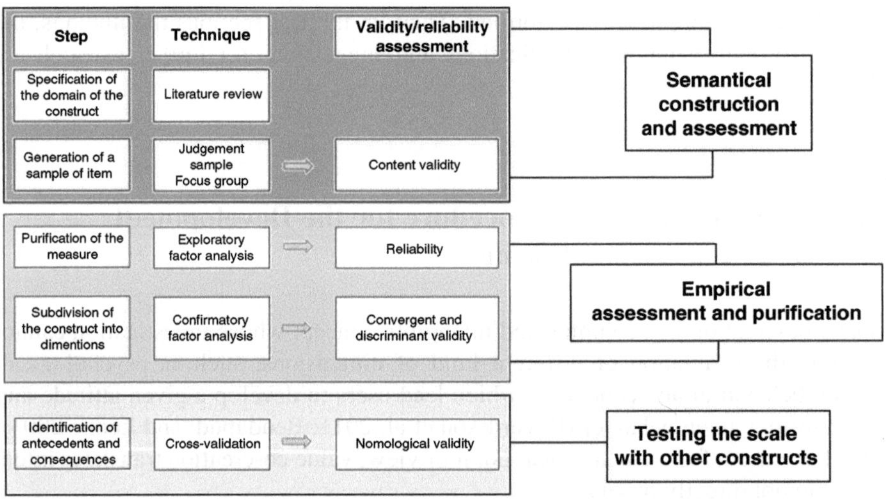

Fig. 8.1 The multi-step process for developing a measurement scale. Elaboration from: Bearden et al. (1993), Churchill (1979), Smith et al. (1996)

[1]To get better results, Churchill (1979) suggests to measure constructs using multi-item following the tradition of psychometrics. Other scholars have shown that the predictive validity of single-item measures is equivalent to multiple-item measures (Bergkvist and Rossiter 2007). In this study, we agree with Churchill's suggestion.

specification of the domain of the specific construct, aimed at circumscribing it from a semantical point of view and, in the second place, the generation of a sample of items in line with the captured domain. At this level, a set of items for each dimension is identified, in an effort to ensure simultaneously the specificity of every item and the semantical similarity with others.

The second macro-area (*empirical assessment and purification*) consists of evaluating the internal coherence of the instrument (reliability) and its validity (convergent and discriminant) respectively through exploratory and confirmatory factor analysis.

The last macro-area (*testing the scale with other constructs*) includes the assessment of the instrument's nomological validity. Finally, instrument's generalizability is tested.

The remaining sections of the paper analyse each stage in detail. In particular, the first two steps have been herein performed; additionally, regarding the last step this work provides useful suggestions for the semantical identification of antecedents and consequences of value co-creation.

8.3 Semantical Construction and Assessment of the Construct

8.3.1 Step 1: Specifying Customer Value Co-creation Domain

The specification of the domain of the construct includes a preliminary theoretical research aimed at performing an in-depth semantical analysis of the concept which in turn should lead to its better understanding and at selecting what should be included and what should be excluded (in terms of relevant dimensions) in its further measurement and in item generation. For this reason, the following sub-paragraphs briefly discuss the results of literature review on value co-creation and advance a classification of the construct into sub-dimensions.

8.3.1.1 Brief Overview on Value Co-creation

As previously stated, numerous studies within S-D logic framework have been realized to systematize value co-creation. However, both theoretical (Neghina et al. 2014; Vargo and Akaka 2012; Wieland et al. 2012; Gummesson and Mele 2010) and empirical works (Yi and Gong 2013; Gustaffson et al. 2012; Randall et al. 2011; Xie et al. 2008) fail to adequately clarify the latent dimensions of the construct, evidencing the existence of a gap.

To date, few studies have explored the nature of dimensionality of customer value co-creation behaviour, leaving its precise composition unclear. Some scholars use a multidimensional approach; in these studies, customer value co-creation behaviour is considered to consist of many components (Bettencourt 1997; Bove et al. 2008; Groth 2005). Other scholars use a unidimensional approach based on single- or multiple-item measures (Cermak et al. 1994; Dellande et al. 2004; Fang et al. 2008). This approach is overly restrictive disregarding the conceptual richness of customer value co-creation behaviour.

In truth, notwithstanding several researchers state their conformity with the theory, over the course of time the gradual departure from Vargo and Lusch's principles and the copious application of the construct to numerous research streams has created further ambiguity and ungeneralizable models.

In particular, among the various contributions seeking to systematize value co-creation, Neghina et al. (2014) identify the sub-dimensions of value co-creation and its antecedents and consequences. However, by espousing *service logic* (Grönroos and Voima 2013), the work ascribes value co-creation only to the interactive moment of delivery (provider-user encounter) and does not follow S-D logic overall conceptualization (in which value is also in co-delivery and post-delivery).

From the perspective of consumer culture theory (CCT), Xie et al. (2008) investigate the underlying dimensions predisposing users to *prosumption* (Toffler 1980) which in turn generate value co-creation. So, the empirical work, which furthermore totally assimilates value co-creation to presumption, focuses only on psychological aspects preceding consumers' involvement.

In consumer behaviour research, instead, Yi and Gong (2013) consider value co-creation as a third order factor through the lens of two theories, customer participation behaviour and customer citizenship behaviour (CPC and CCB, Yi et al. 2011). So, the identification of the measures of customer value co-creation is not devised within the mainstream of S-D Logic; what is more, also in this case there is a focus on the interactive feature of the process.

By introducing a three-variable measurement scale (*connection*, *trust* and *commitment*), Randall et al.'s (2011) study is centered on the relational dimension of value co-creation in line with a customer Relationship Management (CRM) approach (Girishankar 2000; Newell 2001) which takes into account only on social connections involved in the process.

This brief overview highlights that previous research partially explores the relationship between the overall construct and all its semantical facets and tends to adopt a micro-level of analysis (or a too specific standpoint). It follows that there are only a few studies (Chandler and Chen 2015; McColl-Kennedy et al. 2012) that pursue the specific aim of value co-creation measurement, at the same time completely embracing S-D logic and an overall systemic perspective. So it is appropriate to:

1. Clearly identify and measure customers' behavior in co-creating value;
2. Fully validate a comprehensive customer value co-creation behaviour construct;
3. Explore the dimensionality of customer value co-creation behavior.

Therefore, in this study, the definition of value co-creation proposed in the S-D Logic mainstream is espoused, consequently interpreting value co-creation as a comprehensive and multi-stakeholder process that involves, in addition to suppliers, other entities, such as firms, public and private sources and personal actions undertaken by consumers leading to a final multilevel knowledge exchange (Vargo et al. 2008).

Starting from the theoretical work of Tommasetti et al. (2015a), some steps toward a scale measurement model for value co-creation behaviour, strictly adhering to S-D Logic, are herein proposed.

8.3.1.2 Dimensioning Value Co-creation

Tommasetti et al.'s model (Tommasetti et al. 2015a) suggests that customer value co-creation behavior has a hierarchical factor structure. By combining the results of this work and McColl-Kennedy et al.'s (2012) framework, it is possible to hypothesize that value co-creation behaviour is a construct that summarizes a set of eight complex activities.

As depicted in Fig. 8.2, the dimensions recognised are *cerebral activities*, *cooperation*, *information research and collation*, *combination of complementary activities*, *changing habits*, *co-production*, *co-learning* and *connection*.

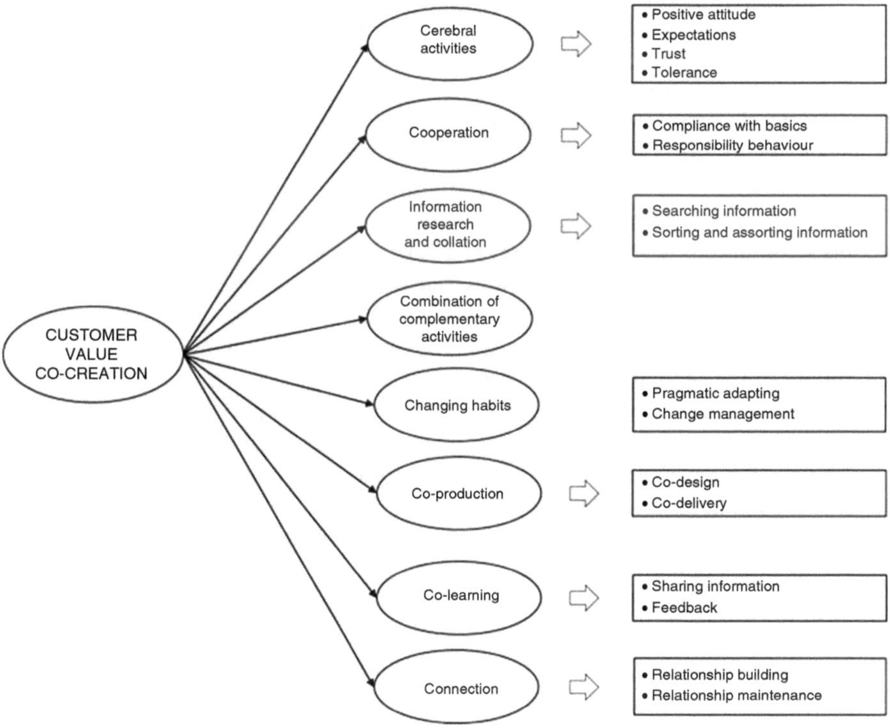

Fig. 8.2 The value co-creation measurement conceptual framework

The first dimension refers to the customers' *cerebral activities* involved in the provision of service. It is composed of three sub-dimensions: (1) positive attitude; (2) compliance with basics; (3) trust; (4) tolerance.

Users' positive attitudes towards suppliers (Bagozzi 1992; Fishbein and Ajzen 1975) is a general (in favourable or unfavourable) mood predisposing customer's to service delivery, an approach impacting on customers' psyche. Expectations related to service (Cardozo 1965), instead, concern consumers' mental attitudes towards their involvement in service delivery; it can be considered as a driver for actor's all-round participation and for continuous improvement of service (Parasuraman et al. 1991). Trust corresponds also to a psychological state involving the intention to accept vulnerability deriving from the uncertainty about the behaviour of others (Rousseau et al. 1998), whereas tolerance is user's capability to sustain all the possible inefficiencies of services (Parasuraman et al. 1991) which do not meet expectations.

The *cooperation* dimension, instead, includes the acceptance and the compliance with provider's guidelines pertaining to the service (Dellande et al. 2004; Bettencourt 1997) in terms of conformity with the instructions provided by service suppliers targeted to the correct execution of tasks. The second sub-dimension of cooperation refers to the consequent responsible behaviour performed by consumers who agree with the rules established (Ennew and Binks 1999) and who have obligations connected with their role of part-time employees (Bowen 1986, p. 373). Cooperation is generally centered on the understanding and sharing of general shared value propositions, routines and institutional arrangements (Vargo and Lusch 2016) with service providers.

A further dimension is the *search for and the collation of information* (McColl-Kennedy et al. 2012; Caridà et al. 2014), consisting of a set of basic informative actions carried out by users and useful to satisfy their needs in order to comprehend service modalities. After the search for information related to service, users sort and assort information (McColl-Kennedy et al. 2012), that is the general organization of the information itself. These activities can be useful to enhance service delivery (Tommasetti et al. 2015b) and to prevent possible inefficacies.

Combination of complementary activities is related to customers' participation in further activities and in events related to the service or in additional services related to basic service and not fundamental for supply but designed to give users some competencies about the service domain. This dimension is borrowed from healthcare (Kremser et al. 2008), in which it represents patient's practice of mixing alternative medicine with the main treatments. Collateral services can be organized by suppliers to strengthen user engagement through the extension of interactive moments of encounter with users.

Changing habits pertains to the modification of consumers' routine and practices caused by service provision. To deal with these alterations and to raise value co-creation, consumers should be able to manage long-term changes deriving from service (pragmatic adapting) and to govern these changes (change management). Pragmatic adapting (McColl-Kennedy et al. 2012) is intended as the stage in which customers accept that service delivery could cause changes in their daily lives and practices. After having approved these modifications, consumers should manage these changes, both in the short and long term (McColl-Kennedy et al. 2012). This phenomenon occurs frequently in service sectors in which fruition is continued and in which actors deal with sacrifices related to consumption.

Co-production represents consumers' engagement in service delivery. It is a central phase of value co-creation, which is both preliminary to service provision (co-design and co- development of a value proposition, Lusch and Vargo 2014) and at the same time contemporary to it (co-delivery).

Lusch and Vargo (2014) describe co-design as consumers' involvement in the phases preceding provision through the joint development of a common value proposition with suppliers, who can grab some suggestions and ideas for implementing a more effective and efficient offering (Prahalad and Ramaswamy 2004; Grönroos and Voima 2012). Co-delivery (Bovaird and Loeffler 2012; Kannan and Chang 2013), instead, refers to actor's active participation in the interactive moment of encounters with provider. It is a process of mutual service provision in which service outcome is co-created in a unique setting and through a one-time experience.

In the stage of *co-learning*, buyers share with other actors of customer's service network information (Tommasetti et al. 2015b) deriving from external sources, e.g. friends, family, other customers, or internal sources, such as previous knowledge and competencies (Vargo and Lusch 2008). This phase consists of sharing knowledge, information and expertise related to the service domain owned by users and produces a circular learning process Payne et al. (2008) in which from (old) resources exchange (new) knowledge is created. The second sub-dimension of co-learning is feedback between users and providers, that is the set of information which consumers (voluntarily or involuntarily) share with suppliers (Groth et al. 2004).

Seeing as value co-creation is indissolubly based on the bidirectional and interactive exchanges between users and providers (Mele and Polese 2011; Randall et al. 2011), the last dimension of the model introduced is *connection*, in other words the concrete relations among all the members involved in the process. There are two distinct relational stages: (1) relationship building, in which customers establish links with suppliers; (2) relationship maintenance, resulting from the strengthening of trust-based links among customers and users which can lead to satisfaction and loyalty in the long run (Palmatier 2008; Barile 2009).

8.3.2 Step 2: Generating a Sample of Items (Content Validity)

The second phase, the generation of a sample of items, consists in the individuation of a set of items for each dimension, seeking to balance the trade-off between the specificity of every single element and the communality with others. This stage can be conducted through various methods, such as experience survey, stimulation of insights or critical incidents (Churchill 1979). In accordance with the purpose of present research to resolve a gap in the conceptual classification of the notion, the first technique, involving a panel of experts on the subject of Service-Dominant logic, is here selected.

Starting from the definition of the main dimensions or components identified in the previous stage, a two-step procedure for item creation has been conducted: (1) item generation; (2) item editing.

In the first place, a first generation of the items for customer value co-creation measurement has been conducted by five researchers, starting from a critical analysis and reelaboration of the dimensions identified in previous. The goal of this phase is to ensure a proper and full semantical coverage of the construct through the elaboration of measurement items (in the form of assertions) which capture all different shades of meaning of value co-creation. Thus, this stage is aimed at obtaining content validity, that is the extent to which the items selected appropriately sample the content domain (Lawshe 1975; Churchill 1979).

The initial pool of items (initially composed of 130 elements), subdivided in the eight dimensions discussed above, has undergone a gradual process of sharpening. Therefore, a different judgement sample has been asked to link each item/statement to one of the dimensions. A pretest has been administered to a panel of 10 academics, marketing and business experts familiar with value co-creation literature and, whose task was to read the definitions of the dimensions and then assign to each item the letter corresponding to the relative definition. The selected sample of professors provided suggestions and comments on items appropriateness, clarity and ease of comprehension, evaluating in particular how each item pertains to the relative dimension.

To attain a final list of items resulting from the unanimous judgements of the experts, Krippendorff coefficient, a correlation coefficient (C, Krippendorff 1980) evaluating the level of agreement among independent observers on a subject, has been utilized. In particular, the degree of inter-subjectivity among different recordings made on the same items by different researchers with the same instrument has been estimated. This consistency indicator derives from the ratio of observed discrepancies and observed concurrencies. It is a relative index which ranges from 0 to 1: in case of discrepancy, for the same variable, in the classification of all cases by all analysts, C is equal to 0, whereas in case of total correlation in the classification of all cases, C is equal to 0 (Krippendorff 2004; Losito 1972). In this work, items have been maintained for C values arriving at cutoff of 0.67 (Krippendorff 2004; Wolf 2008). Finally, each statement has been reviewed collectively and then reworded so as to be as precise as possible and to improve or delete ambiguous questions.

Lastly, in the final phase, which is still in progress, 8 economics and management students further evaluated the purified 70 items, rating how well each of them reflects the different dimensions of customer value co-creation behavior. Only items considered clearly representative at least by five subjects (Bearden et al. 2001; Tian et al. 2001; Zaichkowsky 1985) will be maintained.

As a result, 28 items have been excluded from the research, whereas the remaining 42 have been rephrased for clarity. In the final list, each construct is operationalized with about five items; some of the statements are reformulated based on the work of McColl-Kennedy et al.'s (2012) and Tommasetti et al.'s (2015a), whereas the others have been elaborated from the literature review on value co-creation activities.

8.4 Empirical Assessment and Purification

The third and the fourth step of the methodological procedure herein discussed are intended to purify the measurement scale in order to ensure that each item properly balances the trade-off between distinctiveness and specificity in semantically representing the construct (Churchill 1979).

In line with the aim of measurement, which is targeted to assign numerical values to attributes regarding specific objects, there are two kinds of: (1) reliability, an index of internal consistency of a set of items; (2) validity, related to the extent to which the items (more or less) accurately "cover" the construct.

Since the final goal of measurement is to capture the real core of the phenomenon investigated, reliability and validity can be intended in terms of deviation between observed scores (X_O, what is measured) and true score X_T (true variability in the opinion of a series of subjects with regard to the characteristics measured). So, it can be stated that a measure is valid when it is shown to represent exactly the phenomenon of interest and when the gap between X_O and X_T tends to be reduced. In other words, validity is the degree of semantical correspondence between the construct in the real world and the way in which it is measured depending on the specificity of the single research and study context. A measure is reliable, instead, when it is shown to systematically represent the phenomenon of interest in a consistent manner over time. Reliability evaluates the degree of concordance between two comparable measures of the same construct conducted over time and it depends on how much the variation in scores is attributable to random errors (Churchill 1979). Consequently, when a measure is valid it is also reliable but it is not true the converse relation. So reliability is a necessary but not sufficient condition for validity.

However, the problems concerning the measurement exceed the goal of this study and for this reason they will not be treated. For the same reason, the relationship between indicators and constructs is not covered specifying only that when the construct is latent the indicators are reflective, in the other case they are formative (Edwards and Bagozzi 2000).[2]

[2]The main difference can be synthesized in the following statement. When the model is reflective the common cause of item or indicator behavior derives from the latent variable and the causal action moves from the latent variable to the indicators. In a formative model the composite variable (i.e. the construct) summarizes the variation of indicators and the causal action flows from indicators to the construct.

Even so, it is clear that the objective of measurement is to produce X_O scores which approximate X_T as far as possible. This result is connected with the procedures that are utilised "… to develop the measures and the evidence supporting their 'goodness'" (Churchill 1979, p. 65).

8.4.1 Step 3: Purifying the Measure (Reliability)

The final pool of items obtained in the previous stage can be further sharpened through exploratory factor analysis (EFA, Gerbing and Anderson 1988), which can contemplate a phase of data collection based on a second sample of subjects (in this case they could be patients, for example). This technique allows to determine the number of dimensions (factors) underlying the latent construct (value co-creation) and to assess reliability since it reveals the cohesion of a set of items with regard to the macro- construct.

EFA is based on a theoretical model assuming that the measurement items for each of the latent factors are not a priori defined. The analysis is designed to determine the value of the factorial coefficients that describe the relationship (in terms of correlation and covariance) between the items and factors; so, the aggregate factorial coefficients measure the same latent factor. Exploratory factor analysis should start from a support theory and/or a recognized model that provide for an adequate interpretation of the extracted latent factors.

This technique is performed together with other instruments for assessing reliability. Among these, Cronbach's Alpha (Cronbach 1951; Nunnally 1967) is one of the most common coefficients employed to estimate internal consistency of a set of items and the random errors due to the instrument. It derives from the ratio between the variance between the items and the total variance; so, if Alpha is low, the selected sample of items does not fully grasp the concept (the cutoff fixed in literature is 0.6, Nunnally and Bernstein 1994), if it is high the subjects in the sample express a consistent attitude regarding each item in each dimension. Another method is item-to-total correlation, evaluated through a correlation matrix among all the items. The goal is to measure the average correlation, that is the degree of commonality (common core) between items, and the dispersion from the mean which takes into account the degree of variation of the items. If the items all belong to the same domain, then the degree of commonality will be identical for all the elements, with equal average correlation in each column (interrelated responses) which in turn will be equal to the total correlation (Churchill 1979).

8.4.2 Step 4: Subdividing the Construct into Dimensions (Convergent and Discriminant Validity)

The first three steps of Churchill's methodology concern the evaluation of reliability (internal consistency of a set of items) and of an internal kind of validity (the abovementioned content validity which regards the semantical coverage of the construct), whereas the fourth step introduces the issue of construct validity measurement. This is an external kind of validity aimed at relating the items of a certain construct to the items belonging to other constructs.

Construct validity is subdivided into convergent validity and discriminant validity. The first one assesses the (high) correlation of items measuring the same constructs with different techniques or measures; the latter assesses the (low) correlation of items with other items measuring other constructs and/or with different techniques or measures (Churchill 1979). The convergent validity implies that the items used to measure construct should be internally homogeneous or consistent. In other words, items should measure the same thing and therefore they must be correlated. Consistency is a condition necessary, but not sufficient for demonstrating construct validity (Nunnally 1967).

Thus, in this stage confirmatory factor analysis (CFA) is devoted to the assessment of the two kinds of validity through the confirmation of the items purified and of the underlying dimensions obtained with the EFA. So, the items should be administered to a new sample; in this case, this phase is in progress and the survey has been administered to students.

CFA is halfway between the fourth and the fifth step, since it allows researchers to validate the dimensions obtained with EFA both at an intra-level (convergent validity assessing the relationships between the observed variables and their underlying latent constructs) and at an inter-level (discriminant and nomological validity) that permits to test the relationships of the construct with other variables. Researchers, in fact, starting from literature, elaborate a measurement model hypothesizing a specific relationship pattern among the construct and other variables and then tests it statistically.

Specifically, CFA can involve different statistical tests to determine the adequacy of model fit to the data. Among these, Structural Equation Modeling (SEM) is one of the most commonly adopted techniques in marketing literature (Del Bosque and San Martín 2008; Lee and Song 2004; Nowacki 2009). SEM permits researchers to statistically prove multiple relationships among variables measured with multiple items and allows the concurrent estimation of relationships between multiple dependent and independent variables (Gefen et al. 2000), being suitable for managing complex, immaterial and multifaceted construct not directly observable such as value co-creation.

Anderson and Gerbing (1988) suggest a two-step procedure combining CFA—to estimate the measurement model (relating the construct to its measures)—and the evaluation of structure model, which relates the constructs to each other and in which the hypothetical relationships among all of the variables are identified.

8.5 Testing the Scale with Other Constructs: Toward Nomological Validity—A Research Agenda

As discussed before, to establish discriminant validity, it is necessary to assess that a construct is not correlated to other constructs. In the same way, nomological validity is designed to reveal if a given construct acts as expected in relation to other variables through a further administration of the measurement items (for each variable) to a new sample.

In particular, in order to attain a holistic understanding of value co-creation (Vargo and Lusch 2016), it can be identified below the antecedents and the consequent of the construct. Circumscribing the scope of the concept and individuating the drivers and the effects of a successful value co-creation it is useful to address future research on the topic and to develop a nomological network covering central concept. In literature, moreover, antecedents and consequences related to the sphere of co-creation along Service-Dominant logic pathway have not yet been pinpointed.

In line with the aim of this work to take into account customer's point of view and contribution to value co-creation, also the proposal of the antecedents and consequences of the construct derives from the willingness to study the components predisposing the specific activities and interactions performed by users and the consequences that these actions can generate. The model hypothesized is also centered on an overall vision of value co-creation in which consumers play a leading role from the early stages of co-design to the last stages, being in this way privileged subjects of analysis.

In line with the definition herein espoused, which emphasizes the relational and interactive aspect of the generation of value, not limiting it to a one-to-one relation between consumer and provider, the components individuated are connected to the collaborative aspect of the issue and to the sense of belonging on the part of the consumer.

Concerning the antecedents, one can believe that the factors that predispose to co-creation are related to the degree of users' involvement in corporate activities. So, as emerges from extant research, the constructs that are mostly considered to influence co-creation could be *commitment* (Allen and Meyer 1990) and *customer engagement behavior* (Van Doorn et al. 2010), subsequently called CEB.

The theory of commitment (Allen and Meyer 1990, p. 2) denotes the attachment of a consumer towards a given firm (Bansal et al. 2004; Fullerton 2005; Morgan and Hunt 1994). In the field of marketing, other scholars have defined commitment as the connection between two parties who wish to maintain a relationship (Fullerton 2005; Tommasetti et al. 2014) or the "psychological attachment" to the organization (Gruen et al. 2000, p. 37). According to Walter and Ritter (2003) and Randall et al. (2011) it can be assumed that users which are more immersed in the life of an organization, involved in its conduct, or even delved into its corporate culture, are more willing to co-create.

Instead, customer engagement is a psychological state that occurs by virtue of interactive customer experiences with a focal agent/object such as a firm or brand (Brodie et al. 2011). It is thus a further cerebral dimension that simplifies users' will to take part in the delivery. According to Jaakkola and Alexander (2014) and Banyte and Dovaliene (2014), in fact, CEB affects value co-creation through customers' different resource contributions towards the focal firm and other stakeholders that modify or augment the offering itself, or affect other stakeholders' perceptions, knowledge, preferences, expectations or actions towards the firm or its offering.

With regard to the consequences, it can be affirmed that co-creation influences three variables: customer satisfaction, customer loyalty and innovation.

Customer satisfaction is a complex construct that has been widely debated literature as an assessment of the extent to which the supplier could satisfy or surpass the customers' expectations (Levy and Weitz 2007). Thus, the more customers are active participants in the value co-creation process, the more they are satisfied with the service, which fully adapts to their needs. In addition, Vega-Vazquez et al. (2013) assume a positive relation between value co-creation and customer satisfaction, suggesting that service dedicated to personal care should foster the customers' active participation in the value creation process.

Instead, Rajah et al. (2008) state that buyers' participation, which generates a unique value, can elevate a transaction to a relation-building experience which leads to loyalty: in doing so, through the joint collaboration one can move from simply delivery to an experience, from the asymmetrical relationship between users and providers to a constant long-term relationship based on trust.

Finally, since from the synergic exchange between customers and suppliers it is produced new knowledge (and therefore also new products, offerings, markets and technologies), the co-creation assumes the nature of creative act, being determinant of the production of innovation (Vargo 2013) and "co-innovation" (Romero and Molina 2011, 519 p. 2). In fact, in a network of organizations which involves their customers in joint value creation process, from collaboration among all organizational members, both internal and external, a unique and irreproducible competitive advantage which results in innovation can be generated (Ciasullo et al. 2016; Greer et al. 2016; Cosimato and Troisi 2015).

In summary, as depicted in Fig. 8.3, one can suppose that the all-round involvement of consumers in the entire corporate processes (CEB) and a high sense of belonging to the firm supplier (commitment) encourage value co-creation. In this way, the subjects to produce a combined value that not only creates innovative solutions, but that in the short term gives birth to satisfaction and in the long run customer loyalty.

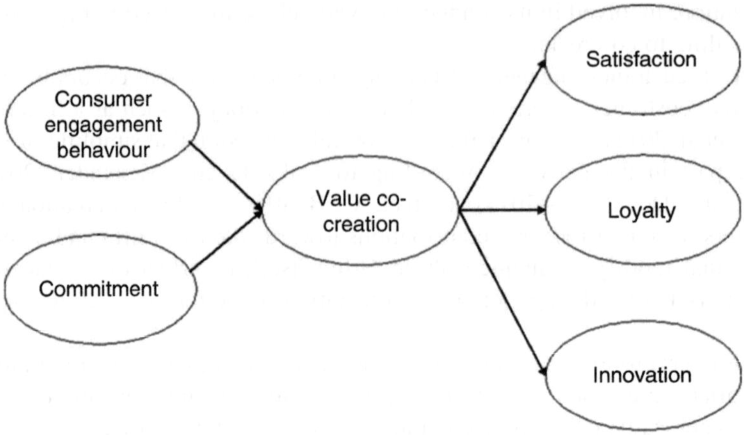

Fig. 8.3 A hypothesis of research model including antecedents and consequences of value co-creation

8.6 Concluding Remarks

The present paper identifies and critically and semantically discusses of the activities composing value co-creation behaviour. In particular, the stages of the multi-level methodology proposed by Churchill (1979)—commonly adopted and in some cases reelaborated in literature—are described step by step and applied to the identification of the main dimensions, antecedents and consequences of value co-creation. In detail, each dimension of the proposed theoretical framework has been semantically defined through the addition of elements or sub-dimension, taking into account several works on customer's behaviour and value co-creation measurement. From the experience survey based on the judgement sample, a final set of 42 items covering 8 dimensions and 16 sub-dimensions has been obtained.

So, this article advances an analytical framework based on S-D logic for an in-depth study of value co-creation and prepares the ground for further measurement of the construct.

The implications of the development of a value co-creation measurement scale are both theoretical and managerial (Greer et al. 2016).

Regarding the first aspect, the holistic view embraced allows the concrete and measurable interpretation of value co-creation as a common denominator between cerebral, emotional, cognitive and social elements—of consumers as well as of suppliers—involved in service delivery.

This work provides a research agenda addressing further research to a correct measurement of value co-creation. It can represent a valuable starting point both from a theoretical and an empirical standpoint, since it suggests that value co-creation is a second order factor composed of a set of activities intended as a whole of interrelated actions and practices performed by users in an ecosystem view.

Relative to managerial aspects, the operationalization of value co-creation sub-dimensions, drivers and consequences is useful to guide managers to the identification of new modes of delivery and to the implementation of ad hoc strategies and tactics to foster the engagement of users at each level.

Particularly, the identification of customer's actions prior, contemporary and following service delivery can guide managers to better understand how value co-creation, and so overall service quality, can be planned, implemented and then assessed at each stage. By adequately managing the whole process, in fact, managers can monitor the emersion of value co-creation and develop in progress some strategies for enhancing it and for improving the process and the general service quality (through service recovery actions, for example).

Moreover, on the basis of the different kinds of value co-creation practices identified, further research could practically measure them and observe their intensity in order to profile customers and shape the offering on the basis of this segmentation to increase business efficiency through more efficient exchanges.

Nevertheless, the main limitation of this work is related to its preliminary nature. Actually, the items generation herein executed, testing validity content of the model and the identification of antecedents and consequences, is only a starting point for developing a value co-creation scale that has to be empirically validated for testing construct validity and nomological validity.

References

Allen, N. J., & Meyer, J. P. (1990). The measurement and antecedents of affective, continuance and normative commitment to the organization. *Journal of Occupational and Organizational Psychology, 63*(1), 1–18.

Anderson, J. C., & Gerbing, D. W. (1988). Structural equation modeling in practice: A review and recommended two-step approach. *Psychological Bulletin, 103*(3), 411.

Arnould, E. J., Price, L. L., & Malshe, A. (2006). Toward a cultural resource-based theory of the customer. In R. F. Lusch & S. L. Vargo (Eds.), *The service-dominant logic of marketing: Dialog, debate and directions*. Armonk: ME Sharpe.

Bagozzi, R. P. (1980). *Causal models in marketing*. New York: Wiley.

Bagozzi, R. P. (1992). The self-regulation of attitudes, intentions, and behavior. *Social Psychology Quarterly, 55*(2), 178–204.

Bansal, H. S., Irving, P. G., & Taylor, S. F. (2004). A three-component model of customer to service providers. *Journal of the Academy of Marketing Science, 32*(3), 234–250.

Banyte, J., & Dovaliene, A. (2014). Relations between customer engagement into value creation and customer loyalty. *Procedia-Social and Behavioral Sciences, 156*, 484–489.

Barile, S. (2009). *Management sistemico vitale* (Vol. 1). Torino: Giappichelli.

Bearden, W. O., Netemeyer, R., & Mobley, M. F. (1993). *Handbook of marketing scales: Multi-item measures for marketing and consumer behavior research*. Thousand Oaks: Sage Publications.

Bearden, W. O., Hardesty, D. M., & Rose, R. L. (2001). Consumer self-confidence: Refinements in conceptualization and measurement. *The Journal of Consumer Research, 28*(1), 121–134.

Bendapudi, N., & Leone, R. P. (2003). Psychological implications of customer participation in co-production. *Journal of Marketing, 67*(1), 14–28.

Bergkvist, L., & Rossiter, J. R. (2007). The predictive validity of multiple-item versus single-item measures of the same constructs. *Journal of Marketing Research, 44*(2), 175–184.
Bettencourt, L. A. (1997). Customer voluntary performance: Customers as partners in service delivery. *Journal of Retailing, 73*(3), 383–406.
Bovaird, T., & Loeffler, E. (2012). From engagement to co-production: How users and communities contribute to public services. In T. Brandsen & V. Pestoff (Eds.), *New public governance, the third sector and co-production*. London: Routledge.
Bove, L. L., Pervan, S. J., Beatty, S. E., & Shiu, E. (2008). Service worker role in encouraging customer organizational citizenship behaviors. *Journal of Business Research, 62*(7), 698–705.
Bowen, D. E. (1986). Managing customers as human resources in service organizations. *Human Resource Management, 25*(3), 371–383.
Brodie, R. J., Hollebeek, L. D., Jurić, B., & Ilić, A. (2011). Customer engagement: Conceptual domain, fundamental propositions, and implications for research. *Journal of Service Research, 14*(3), 252–271.
Cardozo, R. (1965). An experimental study of customer effort, expectation, and satisfaction. *Journal of Marketing Research, 2*(8), 244–249.
Caridà, A., Colurcio, M., & Melia, M. (2014). Interactive health technologies and value co-creation: The Mayo Clinic experience. In E. Baglieri & U. Karmarkar (Eds.), *Managing consumer services: Factory or theater?* New York: Springer International Publishing.
Cermak, D. S. P., File, K. M., & Prince, R. A. (1994). Customer participation in service specification and delivery. *Journal of Applied Business Research, 10*(2), 90–97.
Chandler, J., & Chen, S. (2015). Prosumer motivations in service experiences. *Journal of Service Theory and Practice, 25*(2), 20–239.
Churchill, G. A., Jr. (1979). A paradigm for developing better measures of marketing constructs. *Journal of Marketing Research, 16*(1), 64–73.
Ciasullo, M. V., Polese, F., Troisi, O., & Carrubbo, L. (2016). How service innovation contributes to co-create value in service networks. In *International Conference on Exploring Services Science* (pp. 170–183). Cham: Springer International Publishing.
Cosimato, S., & Troisi, O. (2015). Green supply chain management: Practices and tools for logistics competitiveness and sustainability. The DHL case study. *The TQM Journal, 27*(2), 256–276.
Crocker, L. M., & Algina, J. (1986). *Introduction to classical and modern test theory*. New York: Holt, Rinehart, and Winston.
Cronbach, L. J. (1951). Coefficient alpha and the internal structure of tests. *Psychometrika, 16*(3), 297–334.
Del Bosque, I. R., & San Martín, H. (2008). Tourist satisfaction a cognitive-affective model. *Annals of Tourism Research, 35*(2), 551–573.
Dellande, S., Gilly, M. C., & Graham, J. L. (2004). Gaining compliance and losing weight: The role of the service provider in health care services. *The Journal of Marketing, 68*(3), 78–91.
Edvardsson, B., Tronvoll, B., & Gruber, T. (2011). Expanding understanding of service exchange and value co-creation: A social construction approach. *Journal of the Academy of Marketing Science, 39*(2), 327–339.
Edwards, J. R., & Bagozzi, R. P. (2000). On the nature and direction of relationships between constructs and measures. *Psychological Methods, 5*, 155–174.
Ennew, C. T., & Binks, M. R. (1999). Impact of participative service relationships on quality, satisfaction and retention: An exploratory study. *Journal of Business Research, 46*(2), 121–132.
Fang, E., Palmatier, R. W., & Evans, K. R. (2008). Influence of customer participation on creating and sharing of new product value. *Journal of the Academy of Marketing Science, 36*(3), 322–336.
Fishbein, M., & Ajzen, I. (1975). *Belief, attitude, intention and behavior: An introduction to theory and research*. Reading: Addison-Wesley.

Fullerton, G. (2005). The service quality–loyalty relationship in retail services: Does commitment matter? *Journal of Retailing and Consumer Services, 12*(2), 99–111.

Gefen, D., Straub, D., & Boudreau, M. C. (2000). Structural equation modeling and regression: Guidelines for research practice. *Communications of the Association for Information Systems, 4*(1), 7.

Gerbing, D. W., & Anderson, J. C. (1988). An updated paradigm for scale development incorporating unidimensionality and its assessment. *Journal of Marketing Research*, 186–192.

Girishankar, S. (2000). Companies want CRM tools to manage business relationships. *Information Week, 17*, 65.

Greer, C. R., Lusch, R. F., & Vargo, S. L. (2016). A service perspective. *Organizational Dynamics, 1*(45), 28–38.

Grönroos, C., & Voima, P. (2012). Making sense of value and value co-creation in service logic. *Journal of the Academy of Marketing Science, 41*(2), 133–150.

Grönroos, C., & Voima, P. (2013). Critical service logic: Making sense of value creation and co-creation. *Journal of the Academy of Marketing Science, 41*(2), 133–150.

Groth, M. (2005). Customers as good soldiers: Examining citizenship behaviors in internet service deliveries. *Journal of Management, 31*(1), 7–27.

Groth, M., Mertens, D. P., & Murphy, R. O. (2004). Customers as good soldiers: Extending organizational citizenship behavior research to the customer domain. In D. L. Turnipseed (Ed.), *Handbook of organizational citizenship behavior* (pp. 411–430). Hauppauge: Nova Science Publishers.

Gruen, T. W., Summers, J. O., & Acito, F. (2000). Relationship marketing activities, commitment, and membership behaviors in professional associations. *Journal of Marketing, 64*(3), 34–49.

Gummesson, E., & Mele, C. (2010). Marketing value co-creation through network interacting and resource integration. *Journal of Business Market Management, 4*(4), 181–197.

Gustaffson, A., Kristensson, P., & Witell, L. (2012). Customer co-creation in service innovation: A matter of communication? *Journal of Service Management, 23*(3), 311–327.

Jaakkola, E., & Alexander, M. (2014). The role of customer engagement behavior in value co-creation: A service system perspective. *Journal of Service Research, 17*(3), 247–261.

Johnson, M. D., & Selnes, F. (2004). Customer portfolio management toward a dynamic theory of exchange relationships. *Journal of Marketing, 68*(2), 1–17.

Kannan, P. K., & Chang, A. M. (2013). *Beyond citizen engagement: Involving the public in codelivering government services*. Washington, DC: IBM Business of Government.

Kremser, T., Evansa, A., Moorea, A., Luxforda, K., Begbieb, S., Bensoussanc, A., Mariglianid, R., & Zorbasa, H. (2008). Use of complementary therapies by Australian women with breast cancer. *The Breast, 17*(4), 387–394.

Krippendorff, K. (1980). *Content analysis*. Beverly Hills: California Sage Publications.

Krippendorff, K. (2004). Reliability in content analysis. *Human Communication Research, 30*(3), 411–433.

Lambert, D. M., & Enz, M. G. (2012). Managing and measuring value co-creation in business-to-business relationships. *Journal of Marketing Management, 28*(13-14), 1588–1625.

Lawshe, C. H. (1975). A quantitative approach to content validity. *Personnel Psychology, 28*(4), 563–575.

Lee, S. Y., & Song, X. Y. (2004). Evaluation of the Bayesian and maximum likelihood approaches in analyzing structural equation models with small sample sizes. *Multivariate Behavioral Research, 39*(4), 653–686.

Levy, M., & Weitz, B. A. (2007). *Retailing management*. New York: McGraw-Hill.

Losito, G. (1972). *Analisi del contenuto e sociologia delle comunicazioni di massa*. Milano: Franco Angeli.

Lusch, R. F., & Vargo, S. L. (2014). *The service-dominant logic of marketing: Dialog, debate, and directions*. London: Routledge.

McColl-Kennedy, J. R., Vargo, S. L., Dagger, T. S., Sweeney, J. C., & Van Kasteren, Y. (2012). Health care customer value cocreation practice styles. *Journal of Service Research, 15*(4), 370–389.

Mele, C., & Polese, F. (2011). Key dimensions of service systems: Interaction in social and technological networks to foster value co-creation. In H. Demirkan, J. Spohrer, & V. Krishna (Eds.), *The science of service systems* (pp. 37–59). Heidelberg: Springer.

Morgan, R. M., & Hunt, S. D. (1994). The commitment-trust theory of relationship marketing. *The Journal of Marketing*, 20–38.

Neghina, C., Caniëls, M. C. J., Bloemer, J. M. M., & Van Birgelen, M. J. H. (2014). Value co-creation in service interactions: Dimensions and antecedents. *Marketing Theory, 15*(2), 221–242.

Newell, F. (2001). *Loyalty. com: Customer relationship management in the new era of Internet marketing*. New York: McGraw-Hill.

Nowacki, M. M. (2009). Quality of visitor attractions, satisfaction, benefits and behavioural intentions of visitors: Verification of a model. *International Journal of Tourism Research, 11* (3), 297–309.

Nunnally, J. C. (1967). *Psycometric theory*. New York: Mc-Graw Hill.

Nunnally, J. C., & Bernstein, I. H. (1994). The assessment of reliability. *Psychometric Theory, 3* (1), 248–292.

Palmatier, R. W. (2008). *Relationship marketing*. Cambridge, MA: Marketing Science Institute.

Parasuraman, A., Berry, L. L., & Zeithaml, V. A. (1991). Refinement and reassessment of the SERVQUAL scale. *Journal of Retailing, 67*(4), 420–450.

Payne, A. F., Storbacka, K., & Frow, P. (2008). Managing the co-creation of value. *Journal of the Academy of Marketing Science, 36*(1), 83–96.

Prahalad, C. K., & Ramaswamy, V. (2004). Co-creation experiences: The next practice in value creation. *Journal of Interactive Marketing, 18*(3), 5–14.

Rajah, E., Marshall, R., & Nam, I. (2008). Relationship glue: Customers and marketers co-creating a purchase experience. *NA-Advances in Consumer Research, 35*, 367–373.

Randall, W. S., Gravier, M. J., & Prybutok, V. R. (2011). Connection, trust, and commitment: Dimensions of co-creation? *Journal of Strategic Marketing, 19*(1), 3–24.

Romero, D., & Molina, A. (2011). Collaborative networked organisations and customer communities: Value co-creation and co-innovation in the networking era. *Production Planning and Control, 22*(5–6), 447–472.

Rousseau, D. M., Sitkin, S. B., Burt, R. S., & Camerer, C. (1998). Not so different after all: A cross-discipline view of trust. *Academy of Management Review, 23*(3), 393–404.

Smith, H. J., Milberg, S. J., & Burke, S. J. (1996). Information privacy: Measuring individuals' concerns about organizational practices. *MIS Quarterly, 20*(2), 167–196.

Tian, K. T., Bearden, W. O., & Hunter, G. L. (2001). Consumers' need for uniqueness: Scale development and validation. *The Journal of Consumer Research, 28*(1), 50–66.

Toffler, A. (1980). *The third wave*. New York: Bantam Books.

Tommasetti, A., Botti, A., Troisi, O., & Vesci, M. (2014). Customer satisfaction, commitment, loyalty ed implicazioni nella governance delle università: una ricerca esplorativa. *Azienda Pubblica, 3*, 219–242.

Tommasetti, A., Troisi, O., & Vesci, M. (2015a). Customer value co-creation: A conceptual measurement model in a service dominant logic perspective. In *Proceedings of Naples Forum on Service*, June 2015, Naples.

Tommasetti, A., Vesci, M., & Troisi, O. (2015b). *The internet of things and value co-creation in a service-dominant logic perspective. Data management in pervasive systems* (pp. 3–18). Berlin: Springer International Publishing.

Van Doorn, J., Lemon, K. N., Mittal, V., Nass, S., Pick, D., Pirner, P., & Verhoef, P. C. (2010). Customer engagement behavior: Theoretical foundations and research directions. *Journal of Service Research, 13*(3), 253–266.

Vargo, S. L. (2013). Service-dominant logic reframes (service) innovation. *Highlights in Service Research, VTT Technical Research Centre of Finland, 6*, 7–11.
Vargo, S. L., & Akaka, M. A. (2012). Value co-creation and service systems (re)formation: A service ecosystems view. *INFORMS Service Science, 4*(3), 207–217.
Vargo, S. L., & Lusch, R. F. (2004). Evolving to a new dominant logic for marketing. *Journal of Marketing, 68*(1), 1–17.
Vargo, S. L., & Lusch, R. F. (2008). Service-dominant logic: Continuing the evolution. *Journal of the Academy of Marketing Science, 36*(1), 1–10.
Vargo, S. L., & Lusch, R. F. (2016). Institutions and axioms: An extension and update of service-dominant logic. *Journal of the Academy of Marketing Science, 44*(1), 5–23.
Vargo, S. L., Maglio, P. P., & Akaka, M. A. (2008). On value and value co-creation: A service systems and service logic perspective. *European Management Journal, 26*(3), 145–152.
Vega-Vazquez, M., Ángeles Revilla-Camacho, M. J., & Cossío-Silva, F. (2013). The value co-creation process as a determinant of customer satisfaction. *Management Decision, 51*(10), 1945–1953.
Walter, A., & Ritter, T. (2003). The influence of adaptations, trust, and commitment on value-creating functions of customer relationships. *Journal of Business and Industrial Marketing, 18* (4/5), 353–365.
Wieland, V., Polese, F., Vargo, S., & Lusch, R. (2012). Toward a service (eco)systems perspective on value creation. *International Journal of Service Science, Management, Engineering and Technology, 3*(3), 12–24.
Wolf, J. (2008). *Research methodology. The nature of supply chain management research: Insights from a content analysis of international supply chain management literature from 1990 to 2006* (pp. 38–82). Frankfurt: Verlag.
Xie, C., Bagozzi, R. P., & Troye, S. V. (2008). Trying to prosume: toward a theory of consumers as co-creators of value. *Journal of the Academy of Marketing Science, 36*(1), 109–122.
Yi, Y., & Gong, T. (2013). Customer value co-creation behavior: Scale development and validation. *Journal of Business Research, 66*(9), 1279–1284.
Yi, Y., Nataraajan, R., & Gong, T. (2011). Customer participation and citizenship behavioral influences on employee performance, satisfaction, commitment, and turnover intention. *Journal of Business Research, 64*(1), 87–95.
Zaichkowsky, J. L. (1985). Measuring the involvement construct. *The Journal of Consumer Research, 12*(3), 341–352.

Chapter 9
Market Formation in the Sharing Economy: Findings and Implications from the Sub-economies of Airbnb

Marianna Sigala

Abstract Research in the sharing economy predominately focuses on issues related to the exchange parties and the sharing platforms, ignoring the secondary market of the numerous entrepreneurs emerging around sharing ecosystems. By conducting an exploratory study, this chapter first identified the secondary market entrepreneurs supporting the Airbnb ecosystem and then, it investigated how they impact the sharing accommodation experiences by categorising their services based on the Porter's value chain model. The study also investigated the ability of these entrepreneurs to shape and form new 'hospitality' markets by categorising their market forming capabilities according to the "learning with the market" framework. Findings reveal that the services provided by these entrepreneurs: are similar to the accommodation services provided in the commercialised hospitality context; and they influence the market practices of the 'trading' actors participating in the Airbnb ecosystem. Consequently, the sub-economies created by the secondary market of these entrepreneurs are shaping and evolving the sharing accommodation market to a commercialised 'authentic' hospitality experience.

Keywords Sharing economy • Airbnb • Sub-economies • Entrepreneurship • Hospitality • Experience • Commercialisation • Authentic

9.1 Introduction

The sharing economy is rapidly being diffused in all industries and the accommodation sector is not an exception (Sigala 2014). Consequently, increasing attention is being paid to the peer-to-peer economy in tourism research (Tussyadiah and

M. Sigala (✉)
School of Management, University of South Australia, North Terrace, Adelaide, SA 5001, Australia
e-mail: Marianna.sigala@unisa.edu.au

S. Barile et al. (eds.), *Social Dynamics in a Systems Perspective*, New Economic Windows, DOI 10.1007/978-3-319-61967-5_9

Pesonen 2015). However, research has solely focused on studying issues related solely to the exchange parties (i.e. accommodation hosts and guests) and the sharing platforms, ignoring a growing number of entrepreneurs that are emerging around sharing ecosystems in order to facilitate the trading actors to easier exchange resources in the sharing economy (e.g. Fradkin 2013; Fradkin et al. 2014). On the other hand, preliminary findings show that the sharing economy is fuelling a plethora of new entrepreneurs, and this entrepreneurialism expands the scope and the scale of several economic systems and economies (Botsman 2014; Koopman et al. 2014; Burtch et al. 2016). Claims also exist that the increasing gap between productivity and employment metrics in the USA (showing a decoupling of economic activity and employment) may be due to the entrepreneurial activity fueled by the sharing economy that is invisible in official labour statistics (Badger 2013). These sharing economy entrepreneurs do something so un-traditional that is yet not defined and measured. They follow un-traditional work patterns and they are driven by unconventional and controversial entrepreneurial motives (i.e. drivers combining at the same time commercial entrepreneurial drivers with genuine and altruistic motives to provide social value). Thus, it is also advocated that the services of these micro-entrepreneurs redefine the nature of the initially authentic sharing economy offerings. However, although the sharing economy is challenging the fundamental institutions of employment, work and entrepreneurship (Sundararajan 2016), the literature has failed to far to provide an understanding of this new type and nature of entrepreneurship as well as of the latter's impact on the value and offerings of the sharing economy.

In this vein, the aims of this study are twofold: (1) identify the type of the entrepreneurs that are emerging around the Airbnb sharing ecosystem; and (2) analyse the services provided by these entrepreneurs in order to better understand how their activity affects the hospitality market and experience that are being formed and provided through the Airbnb sharing ecosystem. Airbnb was selected as a case study for this research, because by being one of the biggest sharing accommodation platforms, its industry impact is substantial (Mahajan 2015). Airbnb currently receives more online bookings than any hotel chain, and it is projected to process almost 100 million bookings beyond 2016, with a 40%–50% growth in accommodation offered per year (Huston 2015). Recently, Airbnb also introduced a new B2B functionality allowing companies to book Airbnb listings for their employees. Entrepreneurs operating around the Airbnb sharing ecosystem were found by conducting a major internet search. The websites of these entrepreneurs were analysed for investigating their business models and services. Findings reveal that they basically support Airbnb hosts by outsourcing them (accommodation) management services that are primarily found in the traditional hospitality industry. Consequently, by adapting and transferring traditional accommodation services from the commercialised to the shared accommodation, the sub-economies of these entrepreneurs contribute to the commercialisation of the 'authentic' hospitality experience that Airbnb claims to provide.

9.2 Sharing Economy in the Accomodation Sector: Research Background and Evolution

The sharing economy is widely viewed as a network of connected individuals, communities, and/or organizations that utilizes an online platform to facilitate exchanges, interactions, and experiences for enabling a diversity of exchanges (i.e. lending, renting, swapping, gifting, bartering, sharing, etc.) (Botsman and Rogers 2011). However, although this definition highlights the development of the sharing economy around an ecosystem of various actors interacting and exchanging resources for co-creating value, current research has failed to examine the totality of the actors operating in the sharing economy and specifically, the new entrepreneurs that are emerging around sharing ecosystems and they are contributing with their practices to the restructuring, redefinition and reformation of the traditional economic value systems and chains. Indeed, past research on the sharing economy focuses primarily on the exchanging actors and the sharing platforms (e.g. Fradkin 2013; Fradkin et al. 2014) as well as the socio-economic impacts of the exchanges on communities and employment (Greenwood and Agarwal 2015; Zervas et al. 2015).

Similarly, previous research about the sharing economy in the accommodation sector has examined: the adoption, motivation and values/benefits that hosts and guests derive from trading hospitality services (e.g. Stern 2014; Ikkala and Lampinen 2014); the business model, branding, the technological systems and functionality of the sharing platform like Airbnb (e.g. Fradkin et al. 2014; Yannopoulou et al. 2013); and/or the macro-economic impacts of collaborative exchanges on various socio-economic and cultural issues such as privacy, taxation, legislation, tourism employment, hotel industry, communities and social discrimination (Zervas et al. 2015; Neeser et al. 2015; Miller 2014; Zekanovic-Korona and Grzunov 2014; Guttentag 2013; Fang et al. 2015). Research has also focused on the experiences that people strive to get or they are getting in the shared accommodation. Specifically, findings show that the major experiences and benefits sought by Airbnb guests include: authentic hospitality experiences; social interactions; the experience of domesticity, community, sustainability; but also home and hotel amenities (e.g. laundry facilities, wifi) and low prices (Tussyadiah 2015; Guttentag 2013; Tussyadiah and Pesonen 2016; Lane and Woodworth 2016). These studies mainly highlight that the shared accommodation market is different from the traditional accommodation market, as: it seeks different hospitality experiences; and it represents new tourism market, as the latter would not have travelled unless if the "cheap/affordable" shared accommodation option was available.

On the other hand, another stream of studies also shows that an increasing number of Airbnb guests perceive, select and evaluate the shared hospitality experience in a similar way as the one they use for selecting and evaluating a commercialized hospitality product. For example, Guttentag (2013) found that many Airbnb guests' select shared accommodation options by using criteria that are comparable to the ones used by tourists when selecting traditional accommodation (e.g. service quality, reputation, comfort and equipment, location, price and security). Other studies (Möhlmann 2015; Ert et al. 2016; Olson 2013) have also found that in both the Airbnb and the traditional hospitality context, the guest satisfaction and the likelihood to re-book are determined by the following similar factors: functionality, quality and utility of the accommodation services; trust to the host; economic value and considerations. Moreover, instead of interacting, getting to know each other and exchanging authentic experiences and lifestyles, many studies have also shown that the financial benefit (i.e. extra income or money savings) is the predominant motivation for both sides (guests and hosts) to participate in the sharing economy (Guttentag, 2013; Möhlmann 2015; Tussyadiah 2015). Finally, some studies also challenge the previous arguments that the shared accommodation expands rather than substitutes the traditional hospitality market. For example, Guttentag (2013) and Zervas et al. (2015) found that the lower-priced hotels are negatively affected by the growth of the Airbnb bookings, which in turn confirms the disruption caused by the Airbnb to the traditional accommodation sector.

Moreover, the literature about the sharing economy has also evolved towards a new stream of research examining how shared accommodation providers can participate and effectively use sharing platforms for maximizing their 'sales' and economic benefits. For example: Ert et al. (2016) investigated the factors (e.g. competitors' prices, hosts' reviews and trust) affecting the prices of shared accommodation promoted in Airbnb; Hill (2015) analysed how Airbnb hosts can use the pricing tip tool offered by Airbnb for setting competitive prices. Overall, an increasing number of studies stresses that if Airbnb hosts wish to succeed and to be selected by Airbnb users, then the former have to adopt an operational mindset and management practices (e.g. trust building, managing customer reviews, addressing competitors' prices or applying emotional pricing) that are similar to the business decision-making processes and practices adopted by commercial hotels.

Thus overall, research in the shared accommodation shows that demand-pull factors (i.e. guests' preferences, decision-making criteria and behaviour) and supply-push factors (i.e. maximisation of economic benefits, competitiveness, entrepreneurial goals) exercise pressures to the shared accommodation providers to adopt traditional hospitality management practices and mindset. These simultaneous demand-pull and supply-push pressures demand the Airbnb hosts to act as sharing economy entrepreneurs, social media marketers and hospitality providers. However, shared accommodation providers typically lack general business or specific hospitality knowledge, which in turn undermines their 'business performance' in the sharing economy. This has spurred second-order entrepreneurialism within the sharing economy. Particularly numerous entrepreneurs have set-up new companies that outsource management services (e.g. pricing, booking services and management capabilities) to Airbnb accommodation providers to enable them to operate and trade their hospitality services more efficiently and effectively. However, although professional publications increasingly highlight the mushrooming emergence of entrepreneurs around sharing ecosystems, none academic study has examined so far how these start-up companies support but also transform the value chain, the hospitality experience and the market of the shared accommodation sector.

9.3 Entrepreneurship in the Sharing Economy and Its Role on the Economy and Market Formation

The current socio-economic and technological environment is widely recognised to inspire and fuel entrepreneurial activity. Indeed, knowledge assets and technological advances are considered as the strategic assets and enablers of the modern entrepreneurs (Audretsch and Thurik 2001; Thurik 2008), who also differ from traditional entrepreneurs because of their knowledge-intensive and technology-driven process and attitude (Romano et al. 2016). Modern entrepreneurs need to manage three levels of knowledge (domain, organisational and technological knowledge, Malerba 2010) and they are subject to both business risks (market potential) and technology risks (reliability and continuous innovativeness of technology) (Byers et al. 2011). The sharing economy is identified as a major type of entrepreneurship 3.0, a concept developed by Maeyer and Bonne (2015) to refer to the entrepreneurial activity generated and fuelled by the technological advances and the crowdfunding opportunities. Being knowledge-intensive and technology-driven, the success the sharing economy entrepreneurship depends on three major factors (Standing and Mattsson 2016): market opportunity recognition, business model development and technology commercialisation.

The press (Badger 2013; Zumbrun and Sussman 2015) and academia (Sundararajan 2014; Botsman 2014; Koopman et al. 2014; Burtch et al. 2016) have already started looking into the influence of the sharing or gig-economy to boost entrepreneurial activity through flexible employment and micro-entrepreneurship. However, the literature and preliminary findings do not provide conclusive results into the impact of the sharing economy on entrepreneurship, because (Burtch et al. 2016): people providing their labour in sharing economy platforms do not perceive themselves as entrepreneurs, and so a better understanding and definition of the type of entrepreneurship or employment that the sharing economy creates is required; and the theory provides competing but equally compelling arguments that the sharing economy can either increase or decrease entrepreneurial activity.

On the one hand, it is widely argued that entrepreneurship depends on the availability of slack resources (Aggarwal et al. 2012). For example, Uber and Airbnb enables would-be entrepreneurs to set up their own schedules and working patterns while earning stable pay (Hall and Krueger 2015; Swarns 2014); in turn, by exploiting this flexibility for their own benefit, people can then devote resources to ventures without loosing financial security. On the other hand, the literature also argues that un- and under-employment can drive entrepreneurial activity, because people with unacceptable employment options and/or low opportunity costs are motivated to become entrepreneurs as they have excess time and/or can hope for higher economic benefits (Acs and Armington 2006; Fairlie 2002; Storey 1991). Consequently, for people that pursue entrepreneurship as a means of resolving un-employment or under-employment (Fairlie 2002; Storey 1991), the sharing economy will decrease their entrepreneurial activity by providing them alternate employment opportunities. Only one recent study (Burtch et al. 2016) tested these competing theoretical arguments and its findings confirm a dual impact of the sharing economy on entrepreneurship, as initial evidence reveals that the sharing economy jobs may, on average, substitute for lower quality entrepreneurial activity rather than act as a complement to higher quality entrepreneurial activity. In other words, the sharing economy has a different influence on different types of entrepreneurial activity (low vs. high quality).

Technology-driven and knowledge-intensive entrepreneurship are more than ever indispensable for the competitiveness and survival of individuals, organisations and territories alike (Romano et al. 2016). Modern entrepreneurs boost competitiveness in the knowledge economy, because (Romano et al. 2016): they adopt or adapt existing technologies and increase the propensity for incremental innovation; strengthen social capital and innovation networks; spread knowledge and increase the actors' capacity to absorb new knowledge. The innovative, disruptive, and transformative influence of modern entrepreneurs on the economy has also been studied within tourism, whereby Sigala (2015) showed that e-intermediaries exploiting online knowledge resources and/or technological advances have the potential to respond but also form new tourism markets (market-driven vs. driving-market e-intermediaries). By using the 'learning with the market' approach (Storbacka and Nenonen 2011), Sigala identified three learning capabilities through which knowledge-intensive and technology-driven intermediaries engage with tourism markets for shaping and influencing their formation: *network structure* referring to the firms' ability to develop and maintain networks and ties with other market actors with the purpose to exchange resources and co-create value; *market practices* (exchange, normalized and representational practices) referring to the ways and the institutions that support and frame the actors' interactions and resource exchanges; and the *market pictures* representing the actors' interpretation and understanding of the market, which in turn influence their market practices. These capabilities enable the entrepreneurs to initiate change and form (new) markets by: changing the structure and operations of the economic structure and/or value chain; and engaging in (collaborative) sense-making processes that change the actors' mental models/understanding of the markets (i.e. market pictures) and so, their market practices. The role of market structure on service ecosystems transformation and innovation is also in line with current arguments that value co-creation takes place within complex networks that go beyond dyadic resource exchanges (Chowdhury et al. 2016; Brozovic et al. 2016; Vargo and Lusch 2015). The sharing economy also represents a complex ecosystem enabling numerous actors to network and exchange resources in innovative ways, which in turn can significantly lead to disruption and innovation. Moreover, as the sharing economy redefines basic institutions and understandings of basic concepts (e.g. labour, value), these new meanings can also influence the market pictures and so, practices (behaviours) of sharing entrepreneurs that in turn can cause change and transformation.

9.4 Methodology

The study aimed to identify the entrepreneurs that have emerged around the Airbnb sharing ecosystem in order to provide services to Airbnb hosts that can facilitate and support them to provide accommodation services in a more effective and efficient way. An extensive internet search was undertaken for identifying these entrepreneurs and analysing their services. The following keywords were used in google.com and Bing.com search engines: Airbnb; services; management support; Airbnb entrepreneurs. The Airbnb Open (https://airbnbopen.com/)—an annual event organised by Airbnb for gathering and allowing all Airbnb entrepreneurs to meet, network and exchange services—has also been used for identifying Airbnb entrepreneurs. Finally, the methodology also used a snowball technique for identifying Airbnb entrepreneurs; to achieve that the researchers used her personal contacts being Airbnb hosts for naming entrepreneurs that they use, which in turn was asked to provide contacts of any other related entrepreneurial company. Since the topic is new, new entrepreneurs around the Airbnb sharing ecosystem emerge continuously and there is no list or research identifying, studying and categorising them in any way, this study had to adopt an exploratory method for locating these entrepreneurs through a wide internet research and snowballing technique. Once entrepreneurs were identified their websites were studied and/or telephone interviewed if a website was not available for understanding the services they provided. Services were categorised in themes related to management activities and operations, i.e. marketing, operations, cleaning-maintenance, security, legal and/or business consulting, distribution, pricing etc.

9.5 Analysis and Discussion of the Findings

The majority of the identified entrepreneurs operating around the Airbnb ecosystem represents start-ups providing online property management and booking systems that assist Airbnb hosts to: create their listings on Airbnb; screen guests; have an online reservation and booking system. Many other start-ups represent entrepreneurs offering offline services, such as: property decoration and design services; cleaning services; property security services; accounting and management consulting services (e.g. legal and tax services). Other entrepreneurs have developed and/or adapted an existing technology application that Airbnb hosts can use in order to: manage customer reviews; filter reservation requests by filtering guests based on their creditability, profile and reliability; determine pricing strategies; get

Table 9.1 Entrepreneurs supporting the Airbnb sharing ecosystem: classification of their services based on the Porter's value chain model

Firm infrastructure	• Decoration, design and renovation of accommodation properties		
Human Resource Management	• Training services: http://learnairbnb.com/airbnb-hosting-beginners-guide/ www.hostmaker.com • Legal and tax consultancy services: www.sharedeconomycpa.com www.tryzen99.com • Peer support platforms: www.travelerschat.com www.airhostsforum.com		
Technology development	Provision of technology tools—software for managing bookings, check-ins, promotion—listing in sharing platforms: https://www.pillowhomes.com/home?ref=producthunt www.touchstay.com, www.airspruce.me		
Procurement	• Provision of data, research reports and market intelligence in the Airbnb/shared accommodation economy: • www.Airdna.com: provision of analytics for Airbnb http://www.collaborativeconsumption.com: market intelligence in the sharing economy		
Support activities	*Services and examples of start-up entrepreneurs*		
	Primary activities		
	Inbound logistics—Suppliers—procurement	Production—customer service	Marketing—sales—distribution
Services provided by start-up entrepreneurs	• Security and monitoring equipment for patrolling properties Provision of statistics, research and data relating to the sharing economy	• Laundry, cleaning services • Security services • Property management services • Concierge services (e.g. maps, destination information provision services to guests) • Key handling services	• Booking – sales systems • Pricing systems—Revenue Management systems • Marketing services (e.g. development of the listing of a property, photographs, content creation)
Examples of start-up entrepreneurs	• www.RemoteLock: locking systems • *Security and monitoring systems:* http://au.igloohome.co/airbnb/	• Cleaning, key exchange, property listing: www.MadeComfy.com.au • Airbnb property management services: www.BeyondaRoom.com • Key exchange and front office/concierge services: www.KeyCafe.com Cleaning and key delivery services: www.proprly.com: • Airbnb full property management services (listing, pricing, cleaning, maintenance, guests screening etc.): https://	• Pricing services: • www.Everbooked.com • https://beyondpricing.com/?ref=producthunt http://priceonomics.com/data-services/ www.pricemethod.com • Welcome book for Airbnb guests: www.Touchstay.com

(continued)

Table 9.1 (continued)

		www.guestready.com/en-uk/london Airbnb cleaning services: http://guestprep.com/ Airbnb property management services: https://www.guesty.com/?ref=producthunt Concierge services: http://citytrip.com/?ref=producthunt	

access to educational services about the sharing economy; participate in peer-to-peer networks for exchanging knowledge and best practices with other Airbnb hosts and/or sharing economy entrepreneurs. Table 9.1 categorises the identified entrepreneurs by using the traditional value chain model (Porter 1985). Overall, these entrepreneurs provide a great variety of management services that span all the value creation stages of a value chain. By outsourcing these services to Airbnb hosts, the latter are enabled to produce a hospitality product that is very similar to the commercialised hospitality product rather than a genuine hospitality experience.

In other words, by engaging in this co-creation ecosystem, Airbnb hosts are enabled to commercialise and professionalise the accommodation services that they provide through Airbnb, which traditionally have been promoted in the literature and perceived by the citizens' eyes as authentic and genuine hospitality experiences. Consequently, this creates a head-to-head competition between Airbnb hosts and professional hoteliers, which in turn questions the arguments supporting that the Airbnb serves and targets a totally different tourism market. The latter also reinforces the arguments and pressures to regulate the Airbnb sharing ecosystem in order to create a more equal and fair playing marketplaces for both the traditional hotels and the Airbnb accommodation providers (e.g. by taxing and requiring Airbnb hosts to acquire similar accreditation requirements as traditional hotels). From a competitive point of view, traditional hotels would also need to revisit their strategy and value proposition in order to find a way to differentiate themselves from the commercialised sharing economy accommodation product.

Finally, these entrepreneurs are creating several sub-economies of service exchanges around the Airbnb sharing platform that expand the scope and the scale of the Airbnb impact on economic systems, value chains and employment statistics as well. In this vein, studies investigating the impact of Airbnb on destinations would need to adapt a macro level of analysis in order to capture these multiplier economic and employment effects of Airbnb. Overall, these entrepreneurs redefine both the offering of the sharing accommodation sector as well as the structure and the nature of the economic value system of accommodation provision.

For providing their services and value proposition, these entrepreneurs rely on the following types of knowledge (Malerba 2010):

- Domain knowledge: know-how of accommodation management services, expertise in the accommodation sector and in the shared economy domain
- Organisational knowledge: knowledge about business management and service outsourcing
- Technological knowledge: adaption of technological tools/systems/solutions, regulations, institutions etc. in the shared economy domain

The sustainability of the business model and value proposition of these entrepreneurs heavily depends on both business risks (market potential) and technology risks (Byers et al. 2011). For example, some of the entrepreneurs who were identified 1–2 years ago when the interest search was conducted (Table 9.1) may not exist anymore. As any start-up company, these success of these entrepreneurs heavily depends on the attractiveness, appeal and value proposition of their business model as well as the up-take of the sharing economy in general. All these represent business risks relating to the market potential of the shared accommodation market, the ability of the entrepreneurs to provide a solution that effectively responds to its needs, but also to shapes them. In other words, the success of these entrepreneurs also depends on the adoption of the sharing economy in general, and so, their proactive practices to institutionalise this type of shared accommodation as a 'normal' practice in the market can also influence their successes. Technology risks relate to the success of the technological or business propositions of these entrepreneurs to provide a solution that 'solves' the management issues of Airbnb hosts and enables them to do their work in a more efficient and effective way. To overcome technology risks, the technological solutions and management services provided by these entrepreneurs need to be customised to: the needs and profile of the Airbnb hosts (e.g. micro-entrepreneurs of a tiny business scale without business experience, time and knowledge resources); and the institutional context of the sharing economy (e.g. socio-cultural and legal environment). Overall, in order to survive but also become competitive, the entrepreneurs need to address both the business and technological risks by being adopting a re-active and pro-active approach in responding or shaping these market and technological trends. Consequently, entrepreneurship fuelled within the sharing economy definitely requires new skills and competencies that entrepreneurs did not have to possess some decades ago.

Table 9.2 provides another theoretical lence for interpreting the roles that these entrepreneurs adopt for forming and changing the shared accommodation market and by doing this, ensuring the sustainability of their value proposition. By adopting the 'learning with the market' capabilities framework, Table 9.2 explains how the entrepreneurs of the Airbnb sub-economies engage with and participate in the shared accommodation market for shaping its form. Examples of entrepreneurs that have developed each market formation capability are also provided.

Specifically, several entrepreneurs have created and provide an online marketplace/directory that enables the various actors participating in the shared economy (i.e. consumers, trading partners, outsourcers, researchers etc.) to: search and find a shared economy platform and trading actors; and list, promote and exchange their

Table 9.2 "Learning with the market" capabilities and the role of the entrepreneurs surrounding the Airbnb's sharing ecosystem in changing and forming the shared accommodation market

"learning with the market" capabilities enabling entrepreneurs to form markets	Examples of entrepreneurs and their roles in forming and changing the shared accommodation market
Network structure	Entrepreneurs providing a shared economy marketplace/directory enabling the sharing economy actors to link with each other, network and form ties in an open, transparent and fluid way: http://www.shareable.net/ www.peers.org https://www.peerby.com/ www.compareandshare.com/ http://meshing.it/companies http://www.thepeoplewhoshare.com/sharing-economy-guide/ https://www.justpark.com/creative/sharing-economy-index/ http://www.collaborativeconsumption.com/directory/ http://www.collaborativefinance.org/sharing-economy/
Market practices *Exchange practices*	Entrepreneurs educating and making the accommodation hosts and/or guests ready to – Set up prices in Airbnb – List and promote their offerings in Airbnb – Collect and interpret relevant information for engaging in the shared economy – Interpret and follow the legislation regulating the shared economy (taxation, registration, license etc.) – Identify and use shared accommodation platforms for finding, and booking a shared accommodation – Form peer groups of Airbnb hosts and exchange knowledge, exercise lobbying power
Normalized practices	Entrepreneurs engaging in normalized practices that aim to (www.thepeoplewhoshare.com; www.travelerschat.com www.airhostsforum.com): – Influence policy formation and regulations – Develop a culture of 'sharing' and/or sustainable lifestyle – Develop, adapt and provide technological solutions – Enable Airbnb hosts and guests to build trust amongst each other
Representational practices	Entrepreneurs engaging in practices that influence the image of the sharing economy (www.learnairbnb.com; www.airdna.com): – Collection and dissemination of research and data about the sharing economy

(continued)

Table 9.2 (continued)

"learning with the market" capabilities enabling entrepreneurs to form markets	Examples of entrepreneurs and their roles in forming and changing the shared accommodation market
	– Editing and publication of press releases and conferences in the sharing economy
Market pictures: How actors interpret the market network	Entrepreneurs enabling the market actors to participate in: educational seminars and workshops about the sharing economy; peer to peer learning networks and associations; diffusion of a specific terminology and ideology around the sharing economy (www.learnairbnb.com; www.thepeoplewhoshare.com; www.travelerschat.com www.airhostsforum.com)

services. By enabling the various actors of the sharing economy to link with each other, form ties and connect in various ways, these entrepreneurs shape the shared economy by influencing its network structure by making it more open, transparent, flexible and fluid (i.e. actors can identify, evaluate and participate in shared ecosystems in a 'plug-and-play mode). Entrepreneurs also engage in practices influence the norms and institutions (e.g. technology standards, social norms and values, regulations) governing and shaping the shared accommodation market. Finally, many of the practices (seminars, workshops, peer-to-peer exchange networks) developed by these entrepreneurs also influence the mental schemas (e.g. assumptions, ideas and dominating logic) of the actors participating in the shared economy, which in turn shape the way actors engage and co-create value in the shared economy. It also becomes evident that many entrepreneurs engage in more than one type of practice, which highlights that entrepreneurs possess and develop more than one learning capability for engaging with and shaping the sharing economy market. In this vein, future studies would be interesting to examine whether there is a relation between the learning capabilities and: (1) the 'competitiveness', long-term 'sustainability' and performance of the business model of these entrepreneurs; and (2) the evolution and shape of the sharing economy in terms of its institutionalization, routinization and/or integration within traditional value systems and chains.

9.6 Conclusions and Implications

Research in the sharing economy has primarily focused on the exchange actors and platforms, ignoring the various entrepreneurs emerging around sharing economy ecosystems and aiming to empower and enable actors to participate in the sharing economy. This study aimed to identify and study the value proposition and services provided by the numerous types of entrepreneurs that emerge around the Airbnb ecosystem with the purpose to provide services to Airbnb hosts that enable them to engage, shape and redefine the shared accommodation sector.

However, the findings of this study provide evidence that an increasing number of Airbnb entrepreneurs enable and facilitate citizens to 'become' professional accommodation entrepreneurs that can provide a service and a hospitality experience that is very similar to the commercialised hotel experience. Moreover, many of these entrepreneurs engage in market practices that influence the shape and evolution of the shared accommodation market. By doing this, the entrepreneurs do not only ensure the long-term sustainability of their business model and value proposition, but they also support the institutionalisation, routinisation and ultimately the integration of the shared accommodation sector with the traditional accommodation economy. Airbnb has been proposed as an emerging hospitality phenomenon called network hospitality and defined as (Molz 2013, p. 216) the way people "*...connect to one another using online networking systems, as well as to the kinds of relationships they perform when they meet each other offline and face to face*". Airbnb represents network hospitality, as it enables a hybrid marketplace combining three domains of hospitality (namely private, commercial and social). According to this study, the sub-economies of service exchanges enabled by these entrepreneurs empower more private micro-entrepreneurs to become professional hoteliers and participate in the shared accommodation sector. However, by doing this, they also weaken the social aspect of the networked hospitality (less authentic interactions and more professional accommodation services provided for a price), while also strengthen its commercial dimension (provision of commercialised professional accommodation services and experiences for a price).

As all studies, this research has several limitations. The study represents an exploratory research focusing only on one domain of the sharing economy (i.e. accommodation) and one sharing platform (Airbnb). This study also provides the foundation but also the directions for conducting future research. For example, future studies can further examine the impact of these entrepreneurs on: motivating and empowering more and more citizens to become 'professional hoteliers' and participate in the commercialised accommodation sector by buying properties and outsourcing all their management services to sub-economies; inspiring micro-entrepreneurship and creating employment opportunities and jobs for unemployed and/or under-employed; the evolution and the shaping of the shared accommodation experience; and the behaviour, perceptions and satisfaction of Airbnb guests. To achieve that, future research would have to challenge our conceptualisation and definition of employment, entrepreneurship, working patterns as well as models measuring customer satisfaction and experiences.

References

Acs, Z. J., & Armington, C. (2006). *Entrepreneurship, geography, and american economic growth*. Cambridge, MA: Cambridge University Press.

Aggarwal, R., Gopal, R., Gupta, A., & Singh, H. (2012). Putting money where the mouths are: The relation between venture financing and electronic word-of-mouth. *Information Systems Research, 23*(3-Part-2), 976–992.

Audretsch, D. B., & Thurik, A. R. (2001, March). What's new about the new economy? Sources of growth in the managed and entrepreneurial economies. *Industrial and Corporate Change, 10* (1), 267–315.

Badger, A. (2013). The rise of invisible work: Companies like Airbnb and Etsy are redefining what it means to have a "job." Is that good for the economy? http://www.citylab.com/work/2013/10/rise-invisible-work/7412/

Botsman, R. (2014). Sharing is not just for startups. *Harvard Business Review, 92*(3), 23–26.

Botsman, R., & Rogers, R. (2011). *What's mine is yours: How collaborative consumption is changing the way we live*. New York: Harper Collins.

Brozovic, D., Nordin, F., & Kindström, D. (2016). Service flexibility: Conceptualizing value creation in service. *Journal of Service Theory and Practice, 26*(6), 868–888.

Burtch, G., Carnahan, S. & Greenwood, B. (2016). *Can you gig it? An empirical examination of the gig-economy and entrepreneurial activity* (Ross School of Business Working Paper, working paper No. 138)

Byers, H. T., Dorf, C. R., & Nelson, J. A. (2011). *Technology ventures: Management dell'imprenditorialità e dell'innovazione*. Milano: McGraw-Hill.

Chowdhury, I. N., Gruber, T., & Zolkiewski, J. (2016). Every cloud has a silver lining—Exploring the dark side of value co-creation in B2B service networks. *Industrial Marketing Management, 55*, 97–109.

Ert, E., Fleischer, A., & Magen, N. (2016). Trust and reputation in the sharing economy: The role of personal photos in Airbnb. *Tourism Management, 55*, 62–73.

Fairlie, R. W. (2002). Drug dealing and legitimate self-employment. *Journal of Labor Economics, 20*(3), 538–537.

Fang, Y., Qureshi, I., Sun, H., McCole, P., Ramsey, E., & Lim, K. H. (2015). Trust, satisfaction, and online repurchase intention: The moderating role of perceived effectiveness of e-commerce institutional mechanisms. *MIS Quarterly, 38*(2), 407–427.

Fradkin, A. (2013). *Search frictions and the design of online marketplaces* (Tech. rep., Working paper). Stanford University.

Fradkin, A., Grewal, E., Holtz, D., & Pearson, M. (2014). Reporting bias and reciprocity in online reviews: Evidence from field experiments on Airbnb. http://andreyfradkin.com/ [20/3/2015]

Greenwood, B. N., & Agarwal, R. (2015). Matching platforms and HIV incidence: An empirical investigation of race, gender, and socioeconomic status. *Management Science, 62*(8), 2281–2303.

Guttentag, D. (2013). Airbnb: Disruptive innovation and the rise of an informal tourism accommodation sector. *Current Issues in Tourism*, 1–26.

Hall, J. V., & Krueger, A. B. (2015). *An analysis of the labor market for Uber's driver-partners in the United States*. Mimeo: UCLA.

Hill, D. (2015). How much is your spare room worth? *IEEE Spectrum, 52*(9), 32–58.

Huston, C. (2015, August 13). As Airbnb grows, hotel prices expected to drop Market Watch. Marketwatch. www.marketwatch.com/story/as-airbnb-grows-hotel-pricesexpected-to-drop [12/12/2016].

Ikkala, T., & Lampinen, A. (2014). Defining the price of hospitality: Networked hospitality exchange via Airbnb. In *Proceedings of the Companion Publication of the 17th ACM Conference on Computer Supported Cooperative Work & Social Computing*, (pp. 173–176).

Koopman, C., Mitchell, M., & Thierer, A. (2014). *The sharing economy and consumer protection regulation: The case for policy change*. Arlington: Mercatus Center at George Mason University.

Lane, J. & Woodworth, R. M. (2016). The sharing economy checks in: An analysis of Airbnb in the United States. Retrieved February 13, 2016 from http://www.cbrehotels.com/EN/Research/Pages/An-Analysis-of-Airbnb-in-the-United-States.aspx

Maeyer, C., & Bonne, K. (2015). Entrepreneurship 3.0: tools to support new and young companies with their business models. *Journal of Positive Management, 6*(3), 3–15.

Mahajan, N. (2015). Share. Don't own: The sharing economy takes off. Forbes. Accessed July 1, 2016 from http://www.forbesindia.com/article/ckgsb/share.-dont-own-the-sharing-economy-takes-off/39241/1

Malerba, F. (2010). *Knowledge-intensive entrepreneurship and innovation systems. Evidence from Europe*. Routledge: London.

Miller, S. R. (2014). Transferable sharing rights: A theoretical model for regulating Airbnb and the short-term rental market. Available at SSRN 2514178.

Möhlmann, M. (2015). Collaborative consumption: Determinants of satisfaction and the likelihood of using a sharing economy option again. *Journal of Consumer Behaviour, 14*(3), 193–207.

Molz, J. G. (2013). Social networking technologies and the moral economy of alternative tourism: The case of couchsurfing.org. *Annals of Tourism Research, 43*, 210–230.

Neeser, D., Peitz, M., & Stuhler, J. (2015). *Does Airbnb hurt hotel business: Evidence from the Nordic countries* (Master degree paper). Retrieved June 6, 2016, from https://www.researchgate.net/profile/David_Neeser/publication/282151529_Does_Airbnb_Hurt_Hotel_Business_Evidence_from_the_Nordic_Countries/links/5605310e08aea25fce322679.pdf

Olson, K. (2013). National study quantifies reality of the "Sharing Economy" movement. http://www.campbell-mithun.com/678_national-study-quantifies-realityof-the-sharing-economy-movement [10/9/2015]

Porter, M. E. (1985). *Competitive advantage: Creating and sustaining superior performance*. New York: Free Press.

Romano, A., Passiante, G., & Del Vecchio, P. (2016). The technology-driven entrepreneurship in the knowledge economy. In *Creating technology-driven entrepreneurship* (pp. 21–48). London: Palgrave Macmillan.

Sigala, M. (2014). Collaborative commerce in tourism: Implications for research and industry. *Current Issues in Tourism*, 1–10.

Sigala, M. (2015). From demand elasticity to market plasticity: A market approach for developing revenue management strategies in tourism. *Journal of Travel and Tourism Marketing, 32*(7), 812–834.

Standing, S. & Mattsson, J. (2016). Fake it until you make it: Business model conceptualization in digital entrepreneurship. Journal of Strategic Marketing. doi:10.1080/0965254X.2016.1240218

Stern, J. (2014). Airbnb Benefits from social proof theory. Retrieved February 20, 2015, from www.Josephstern.com

Storbacka, K., & Nenonen, S. (2011). Scripting markets: From value propositions to market propositions. *Industrial Marketing Management, 40*, 255–266.

Storey, D. J. (1991). The birth of new firms—Does unemployment matter? A review of the evidence. *Small Business Economics, 3*(3), 167–178.

Sundararajan, A. (2014). Peer-to-peer businesses and the sharing (collaborative) economy: Overview, economic effects and regulatory issues. Available at: http://smallbusiness.house.gov/uploadedfiles/1-15-2014_revised_sundararajan_testimony.pdf

Sundararajan, A. (2016). *Sharing economy*. Cambridge, MA: MIT Press.

Swarns, R. (2014). Freelancers in the 'Gig Economy' find a mix of freedom and uncertainty. *New York Times*, A14.

Thurik, A. R. (2008). *Entrepreneurship, economic growth and policy in emerging economies.* ERIM report series research in management (No. ERS-2008-060-ORG). Erasmus Research Institute of Management (ERIM). Retrieved September 10, 2016, from http://hdl.handle.net/1765/13318

Tussyadiah, I. P. (2015). An exploratory study on drivers and deterrents of collaborative consumption in travel. In I. Tussyadiah & A. Inversini (Eds.), *Information and communication technologies in tourism 2015* (pp. 817–830). Cham: Springer International.

Tussyadiah, I. P., & Pesonen, J. (2015). Impacts of peer-to-peer accommodation use on travel patterns. *Journal of Travel Research*, 1–19.

Tussyadiah, I. P., & Pesonen, J. (2016). Drivers and barriers of peer-to-peer accommodation stay – An exploratory study with American and Finnish travellers. *Current Issues in Tourism, 1368–3500*, 1–18.

Vargo, S. L., & Lusch, R. F. (2015). Institutions and axioms: An extension and update of service-dominant logic. *Journal of the Academy of Marketing Science, 44*(1), 5–23.

Yannopoulou, N., Moufahim, M., & Bian, X. (2013). User-generated brands and social media: Couchsurfing and AirBnb. contemporary management. *Research, 9*(1), 85–90.

Zekanovic-Korona, L., & Grzunov, J. (2014, May). Evaluation of shared digital economy adoption: Case of Airbnb. In *Proceedings of 37th International Convention on Information and Communication Technology, Electronics and Microelectronics (MIPRO)*, 2014 (pp. 1574–1579). IEEE.

Zervas, G., Proserpio, D., & Byers, J. (2015). *The rise of the sharing economy: Estimating the impact of Airbnb on the hotel industry* [Boston University School of Management Research Paper (2013-16)].

Zumbrun, J., & Sussman, A. L. (2015). Proof of a 'gig economy' revolution is hard to find. Despite the hoopla over Uber, Etsy and the sharing economy, Americans are becoming less likely to be self-employed, hold multiple jobs, data show. *Wall Street Journal*, 26 July, 2015. http://www.wsj.com/articles/proof-of-a-gig-economy-revolution-is-hard-to-find-1437932539

Chapter 10
The Performativity of Value Propositions in Shaping a Service Ecosystem: The Case of B-corporations

Cristina Mele, Tiziana Russo-Spena, and Marco Tregua

Abstract The present chapter builds on performativity research and investigates the performative role of value propositions in shaping a service ecosystem. Performativity is used as a central concept in illustrating how actors influence reality through their representations and their practices. In the investigation, Benefit corporations (hereafter B-corporations) serve as the context for empirical research. The work shows how ideas—the pre-constitute aspects of the ecosystem, such as values, norms and meaning—participate in shaping reality (i.e. the ecosystem) through the translation of such ideas into practice. The performative role of value propositions emerges as a bridge between actors' values, aims, and practices in shaping a service ecosystem as well as increasing ecosystem viability.

Keywords Performativity • Value propositions • Service ecosystems • B-corporations

10.1 Introduction

The purpose of business is not to make a profit. (Drucker 1954)

A significant amount of time has passed since Drucker spoke the words above; however, it is only recently that trends in academia and business have begun to emerge with respect to overcome the strictly business mindset.

First, the triple bottom line is used to emphasize the company's commitment to environmental integrity, social equity, and economic prosperity (Chabowski et al. 2011; Crittenden et al. 2011). Environmental integrity refers to ecological issues in terms of natural resources that companies should preserve and not waste; social equity concerns the impact on society of a company's practices in terms of not

C. Mele (✉) • T. Russo-Spena • M. Tregua
Department of Economics, Management, Institutions, University of Naples, Federico II, Naples, Italy
e-mail: cristina.mele@unina.it

S. Barile et al. (eds.), *Social Dynamics in a Systems Perspective*, New Economic Windows, DOI 10.1007/978-3-319-61967-5_10

harming any group of stakeholders; and economic prosperity focuses on the company's financial performance and competitive success.

Second, a new type of company, the 'for-benefit enterprise', which conducts business oriented towards social issues, has appeared (Sabeti 2011). It has been conceptualized as a different way of framing businesses (Peterson 2015) by walking 'a fine line between the institutional spheres of the business and charity sectors' (Battilana and Lee 2014: p. 409). For-benefit enterprises are a new class of organization that have led to the emergence of the so-called 'fourth sector'—i.e., a new ecosystem (Sabeti 2011). Although this sector is not clearly defined, the characteristic feature of 'fourth-sector organizations' is that they integrate social and environmental aims with business approaches. Some of these organizations go even further, embodying features such as inclusive governance, transparent reporting, fair compensation, environmental responsibility, community service, and contribution of profits to the common good. By pursuing these aims, they envision a different ecosystem.

Third, studies into service-dominant (S-D) logic (Frow et al. 2014; Lusch and Vargo 2015) have recently focused on the service ecosystem—i.e., the entangled system of actors (Vargo and Lusch 2011). This complex system is characterized by mutual value propositions and service provision, and governed by socially constructed institutions (Vargo and Lusch 2016). Wieland et al. (2012) outlined that for each instance of resource integration, service provision, and value creation, the nature of the ecosystem changes and thus the context for the next iteration and determination of value creation changes also.

According to recent research (Barile et al. 2016; Lusch et al. 2016), understanding what conditions enable the formation of service ecosystem is important for both researchers and managers. Taillard et al. (2016) offered a fresh perspective on service ecosystem formation as a process of emergence in which the development of shared intentions enables collective agency whose specific conditions result from, and foster, interdependence among actors. However, while their study explained unplanned emergence, more research is needed on the role of actors' values and activities in actively pursuing the development of a new ecosystem.

The current study aims to understand the contributions of actors' values, aims, and practices in shaping a service ecosystem, as in the case of for-benefit enterprises. Recent work has provided some evidence on how a service ecosystem can be designed in accordance with norms and values in society (Enquist et al. 2006; Edvardsson et al. 2011, 2013) by recognizing that different actors interact via value proposals (Edvardsson et al. 2013); however, how such value propositions contribute to shaping the service ecosystem remains largely unexplored.

In addition, actors' practices seem to play an important role in co-creating service ecosystems. According to S-D logic, actors collaborate by developing social practices. Such collaboration makes these actors more interdependent on resources and values and enhances system viability (Barile and Polese 2010; Lusch and Vargo 2014). Although it is recognized that actors' agency allows them to take actions that shape the ecosystem, the relationship between value propositions and service ecosystems remains vague.

By moving from the definition of value proposition as 'how an actor co-proposes to affect another actors' (Lusch and Vargo 2014, p. 72), this work addresses the relationship between value propositions and a service ecosystem in the context of actors' values and practices, and asks, what is the role of value propositions in shaping a service ecosystem?

In order to answer this research question, the present work builds on performativity research (Kjellberg and Helgesson 2006) and investigates the performative role of value propositions as a bridge between actors' values, aims, and practices in shaping a service ecosystem. Performativity emerges as a central concept in illustrating 'how actors influence reality' (Storbacka et al. 2012, p. 69) through their representations and their practices. Main contribution stays in addressing how ideas—the pre-constitute aspects of the ecosystem, such as values, norms and meaning—participate in shaping reality (i.e. the ecosystem) through the translation of such ideas into practice. In the investigation, Benefit corporations (hereafter B-corporations) serve as the context for empirical research.

The remainder of the paper is organized as follows: firstly, the authors conduct a literature review of service ecosystems, value propositions, and performativity Secondly, the research method and findings are presented, and finally, the paper ends with a discussion and the main implications.

10.2 Literature Review

10.2.1 Service Ecosystems

S-D logic (Lusch and Vargo 2014; Lusch et al. 2016) entails development of the concept of service ecosystems, which are distinct from business ecosystems (Iansiti and Levien 2004). In using the concept of service ecosystems, S-D logic aims to highlight that, similarly to biological ecosystems, business networks are characterized by a large number of loosely interconnected service systems that depend one another for mutual effectiveness and survival. More specifically, Vargo and Lusch (2011) defined the service ecosystem as 'relatively self-contained, self-adjusting systems of resource integrating actors connected by shared institutional logic and mutual value creation through service exchange' (p. 31).

The service ecosystems metaphor denotes actors' interactions, as well as value flows. Each service system (or actor) is able to improve the state of another system by participating in sharing or applying resources and improving its own state by integrating other resources (Vargo and Lusch 2011, 2012). From the service ecosystem view, each actor is a beneficiary of the value while also acting as a provider of resources for mutual value creation (Vargo and Lusch 2012). The emphasis is on the dynamic configurations of actors and their structures, which lead to increased accessness and resourceness, thereby sustaining value-creation processes (Vargo and Lusch 2011, 2014) and ecosystem viability (Barile and Polese 2010; Vargo and Akaka 2012; Wieland et al. 2012). The various interactions taking place in service ecosystems generate different level and norms and meanings develop from each level (Greer et al. 2016) The actions and interactions among actors continuously sustain and reproduce the system by socially constructing institutional logic or mental models that, in turn, influence activities and exchanges and contribute to creating structures designed to shape the service ecosystem and increase viability (Vargo and Akaka 2012; Wieland et al. 2012). In such a complex setting, institutions are key elements of focus as they favour value co-creation through a network of relations that encompass resource integration and service exchange (Edvardsson et al. 2014; Vargo and Lusch 2016).

Institutions influence the ways in which the activities of resource integrators are coordinated and adjusted to each other, and act as a coordinating link that impacts value co-creation efforts (Edvardsson et al. 2014; Vargo et al. 2015). Vargo and Lusch (2016) further explored the role of institutions to update the concept of service ecosystems by addressing networks 'as resource-integrating, service-exchanging actors that constrain and coordinate themselves through institutions and institutional arrangements' (p. 6). Institutions (i.e., various types of routinized, coordinating mechanisms) and institutional arrangements (i.e., the assemblages of interdependent institutions) become essential to understanding value co-creation (Vargo and Lusch 2016).

According to recent calls (Barile et al. 2016; Lusch et al. 2016), understanding the conditions that enable the development of service formation is important for both researchers and managers. Meynhardt et al. (2016) stressed that institutional changes are expected to influence the viability of a service ecosystem in the long run, and represent the dynamic force of ecosystem evolution. Similarly, Koskela-Huotari et al. (2016, p. 7) recently noted that 'actors simultaneously break, make and maintain the institutionalized rules of resource integration' in reconfiguring a service ecosystem. Taillard et al. (2016, p. 2972) addressed service ecosystem formation 'as an emergent process in which individual and collective agency, together with the institutional arrangements of the social system in which they operate, are mutually constitutive entities of that system'.

In relation to investigating the process of the emergence of service ecosystems via shared intentions, the role of value propositions remains vague; thus, the focus must move from unplanned emergence to active pursuit.

10.2.2 Value Propositions

The discussion around S-D logic and service ecosystems (Wieland et al. 2012; Lusch and Vargo 2014; Barile et al. 2016) has led to growing interest in redefining the value propositions concept (Frow et al. 2014), mainly around key issues including the role of resources, the actors' interactions, and the development of practices.

In their seminal article published in 2004, Vargo and Lusch conceptualized value propositions as different combinations of resources that companies deploy to provide input into customer value creation processes. The 'supplier' cannot deliver value but can only offer a value proposition. It is up to the beneficiary of that value proposition to co-create and experience the value, with the value proposition setting expectations of value-in-use (Lusch and Vargo 2014).

This view has also been expanded to include a description of value proposition as reciprocal and mutual promises not just between two parties, but among multiple actors (Vargo and Lusch 2009). Gummesson (2008) claimed that balance centricity epitomizes the network-based stakeholder approach to value creation, while Lusch and Vargo (2014) recognized a multiplicity of actors as resource integrators tied together in shared systems of exchange service systems. Accordingly, Frow and Payne (2011) explored the development of value proposition in key stakeholder market domains. In their view, stakeholder value propositions provide enhanced opportunity for value co-creation by assisting companies in aligning values and stabilizing relationships within their value network. Similarly, Storbacka and Nenonen (2011) proposed the term 'market proposition' to capture the unifying nature of the market and value-creation process.

By adopting a wider perspective, Payne and Frow (2014a) emphasized the service ecosystem as the context for building value propositions, describing how value proposals contribute to the wellbeing of the service ecosystem through the dynamic process of resource sharing and shaping. Here, the focus is on value propositions as a means by which to envision beneficial outcomes (or avoid disruptive ones), which can only be achieved through collaboration in an actor-to-actor context, with the value proposition setting out the potential opportunities for all actors within the service ecosystem. Reciprocal and co-created value propositions are identified through three stages: value propositions to customers, value propositions to key actors, and value propositions within the service ecosystem (Payne and Frow 2014b).

Lusch and Vargo (2014, p. 188) stated that all actors, as resource integrators and part of the service ecosystem, engage in creating value propositions. Value propositions can be seen as invitations to participate in value co-creation processes as 'they are appropriately considered narratives of value potential that are co-created among multiple actors, including the provider and beneficiary' (Vargo and Lusch 2016, p. 10). Greer et al. (2016) have recently addressed that value propositions are not thought only for external customers but for all the other stakeholders.

Other contributions have aimed to shed light on the development process of a value proposition, stressing its recursive and interactive nature in a multi-actor context of practices. Korkman et al. (2010) and Kowalkowski (2011) proposed the adoption of a practice-centred perspective to describe how value propositions are co-created through a reciprocal exchange of knowledge among resource-integrating actors in real-life practices. Ballantyne et al. (2011) positioned the idea of reciprocal value propositions as a communication practice that guides resource integration between stakeholders, bringing closer exchange activity, relationship development, and knowledge renewal. Skålén et al. (2015) suggested that value propositions are configurations of three different practices: (1) provision, (2) representational, and (3) management and organization practices. Value propositions are expected to be built and re-built thanks to practices (Skålén et al. 2015).

These recent approaches have articulated a perspective on value propositions that is less theoretically intuitive and more grounded in what firms and other actors do in the context of multiple and interrelated interactions and actions. Practices develop over time to enable actors in an Actor-to-Actor (A2A) network to coordinate their meaning-making, actions, and behaviours for mutual gain through service-for-service exchange (Lusch and Vargo 2014; Russo Spena et al. 2017).

However, in accordance with the research question outlined above, what remains to be understood is how value propositions influence the development of service ecosystems. We suggest that there is a need to take into account the role of performativity.

10.2.3 Performativity

The term 'performativity' was derived by the language philosopher Austin (1962), who used the expression 'performative utterance' to address circumstances in which saying something is doing something. In a similar way, Hall (2000) stated that 'declarations are performative, not constative, because it is by the utterance of the words that the act is performed' (p. 1). Words are performed as actions, which in turn produces a different world (Loxley 2006). In addressing 'a way of doing things with words' scholars see performativity in the following terms:

- 'Linguistic acts do not simply reflect a world but that a speech actually has the power to make a world' (Jackson 2004, p. 2).
- 'Talking together is acting together' (Cavell 2002, p. 33).
- 'Reiterative power of discourse [is able] to produce the phenomena that it regulates and constrains' (Butler 2011, p. xii).

In relation to marketing, Nordgren (2008) recently addressed the performativity of the discourse of value creation by investigating the formation of discourses and their influence on people on an ontological level, along with their use of language and actions.

> Performativity is seen as an intermediary force within the discourse, which through its various discursive practices links people to specific subject positions and influence them to perform certain acts, which agree with these positions. These acts then in turn influence the one, who is performing the act as well as other people. Hence, the performativity of the discourse names and makes people to become subjects in line with the discourse. (Nordgren 2008, p. 116)

Building on actor-network theory (Latour 2005) and on Callon's (1998) notion of performativity, Kjellberg and Helgesson (2006) argued for a need to broaden the discussion on performativity to take into account multiple theoretical aspects, and for the importance of studying performativity in more mundane markets. A performative approach 'allows those who act to define what the world is made of' (Kjellberg and Helgesson 2006, p. 864). Kjellberg and Helgesson view social reality as an ongoing process of creation: 'we stress the emergent character of reality rather than whether or not it exists independently of our perception of it. They assume that social reality is constantly being shaped and reshaped through a recursive process' (Kjellberg and Helgesson 2006, p. 840). Social reality is thus seen as 'a materially heterogeneous relational effect' (Kjellberg and Helgesson 2006, p. 840); i.e., 'it is produced and stabilized in interaction that is simultaneously material and social' (Law and Urry 2004, p. 395). Performativity is conceived as a process through which shared ideas (e.g., about markets) shape the world (i.e., the real market) by affecting actions. Starting from Latour's (2005) notion of the 'chain of translation', Kjellberg and Helgesson (2006) represented the path from ideas to actions (Fig. 10.1); thus, reality is shaped by linking a world of ideas with the concretized world—i.e., the 'world of ideas' is actualized through translation into the 'world out there', and in turn the latter is idealized through translation of this 'world of ideas'. Reality emerges as a process that links the two worlds through making sense and taking action (Kjellberg and Helgesson 2006). From this perspective, scholars have stated that actors shape markets with their everyday practice; markets are indeed outcomes of performative practices (Kjellberg and Helgesson 2006, 2007; Diaz Ruiz 2012).

Within S-D logic (Lusch and Vargo 2014) the performativity issue is linked to markets by addressing the 'commonality of socially constructed and shared institutions—rules of games—such as language, norms and practices' (p. 25). Being shared across structures, institutions affect practices through which actors shape markets.

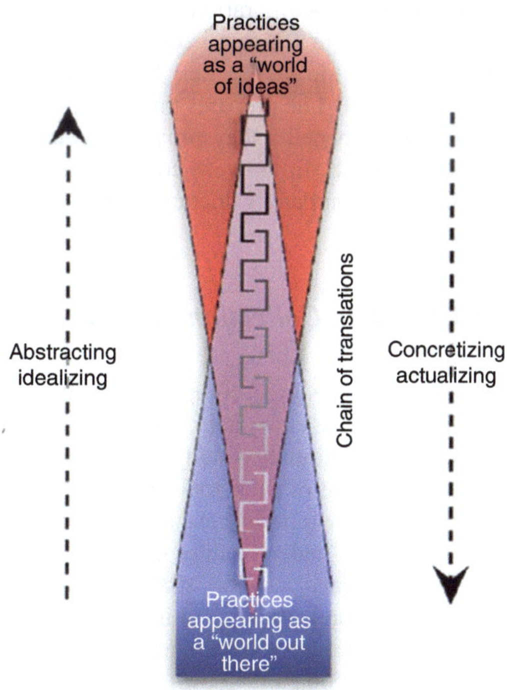

Fig. 10.1 The performativity—the infusion of ideas into the world. Source: Kjellberg and Helgesson (2006, p. 845)

By widening the ideas set forth by Vargo and Lusch (2014) that 'markets are created by the actors through institutionalization, the widespread acceptance, of generalized value propositions (e.g. automobiles) as acceptable solutions to common needs (e.g. personal transportation)' (p. 242), this work addresses the performative nature of value propositions: how actors create institutions and practices that become the glue within the development of service ecosystems.

10.3 Methodology

Multiple case studies were used in this research. The use of qualitative methods allows for deep, detailed, rich data collection that can help to explain complex issues and further develop extant knowledge (Dubois and Gadde 2002; Gummesson 2005).

The research context is formed by a community of B-corporations, a new type of company whose mission is to use the power of business to solve social and environmental problems. Such B-corporations are certified by a nonprofit organization, B-Lab, according to rigorous standards of social and environmental performance, accountability, and transparency. This community includes organizations from more than 60 different industries. The B-corporations community is increasing at an impressive rate, growing from 450 at the end of 2011 to over 26,000 as of March 2017.

A sample of B-corporations was drawn from those companies achieving the best results. Both large and small companies were chosen among those with an 'overall impact in the top 10%'; i.e., firms with the highest performance in terms of sustainability. B-Lab prepares this ranking through a series of measures to compare firms and the benefits they provide to the different stakeholders they are related to. A total of 41 firms were considered; however, three were excluded because access to the company's documentation was not possible.

10.3.1 Data Collection and Analysis

In order to analyse the companies' value propositions and understand their role within the service ecosystem, a two-phase investigation was set up.

Firstly, a preliminary study of the Web-based contexts was conducted, along with observations of the actors and their visible actions in the community (Kozinets 2002). Company reports, case reports, and other related documents were analysed to obtain a pre-understanding of the B-corporations community in terms of the companies' values, mission, strategies, and practices.

Secondly, the authors examined documents published by each organization on the B-corporations website and on the companies' own websites in relation to presenting the companies' own activity and describing their mission and the ways in which they act to change ordinary business into activity that is 'better for the world'. These documents provided an in-depth description of each company's vision/mission, commitment, proposals, and activities. Discourse, objects, and documents (Kozinets 2002) in forums and blogs, along with other descriptive materials were also combined in data setting to create a useful guide to analyse the data. In line with Weber's (1990) suggestion, we used content analysis to classify the textual material and reduce it into more relevant, manageable pieces of data. The analysis began by identifying and quantifying certain words or content in documents to grasp their contextual use. Quantification was conducted to explore usage, rather than to infer meaning, as advised by Hsieh and Shannon (2005). However, the analysis was not only limited to measuring the frequency of specific words or content (i.e., manifest content), but also included a summative approach to *discover* underlying meanings of the words or the content, and used quotations to illustrate issues and phenomena revealed by the investigation (i.e., latent content) (Hsieh and Shannon 2005).

The aim of our data-reduction and -classification process was to identify patterns in the data and generate a local and context-specific understanding of the role of value propositions within the service ecosystem (Gummesson 2005; Piekkari et al. 2010).

The results of the study were illustrated to two senior managers from a B-corporation not included in the investigated cases. A seminar was then organized to discuss the case descriptions and primary implications. This is a good example of a 'member check' procedure (Lincoln and Guba 1985). Taking such steps increases the quality (i.e., the construct validity) of the study in question.

10.4 Findings

The findings show that the B-corporations' value propositions are built on the actors' values and affect their practices by acting on the service ecosystem in a performative way.

10.4.1 Values Within Value Propositions

B-Corporations stay for "Better companies": i.e. Companies that strive to offer something better with the aim to combine business and societal goals. Companies join the B-community to have recognition of and strengthen their commitment to benefiting customers, shareholders, and society.

> For us becoming a B-corporation was a formal acknowledgment of our commitment to the community we serve, our team members, shareholders and the environment. It was another great step we took to ensure that our values and commitment were further engrained in our company's DNA. (documents from Firm N. 28)

> We became a B-corporation to showcase the values that define our business and our lives. (documents from Firm N. 7)

B-corporations share collective values such as responsibility, empathy, empowerment, education, integrity, and sustainability. Shared ethical, social, and environmental principles shape business missions and strategies. Coherent value propositions include not only benefits (arising from the use of services and goods) to customers, but take into consideration value co-creation with a wider audience—namely employees, distributors, suppliers, investors, the community, and also the environment and society. Each group is a beneficiary of the value proposal designed and developed by B-corporations.

Education, environment, and society are the most cited themes. Education is seen in terms of spreading a social and sustainable aim among all actors. B-corporations claim a strong commitment to meeting the challenges of value change and providing associated benefits. They are concerned with their impact on their immediate, as well as more distant, context in order to improve the quality of life.

> B-corporations might turn out to be like civil rights for blacks or voting rights for women—eccentric, unpopular ideas that took hold and changed the world. (Document from Firm N. 13).

> B-corporations' values are aligned with everything we believe in. It's important to educate and allow our consumers to see and understand our business practices as well as our commitment to sustainability. And we only surround ourselves with like-minded individuals and companies—those that believe in giving back and making the world a better place. (B-corporation web site)

Environmental responsibility is a constant issue, and the companies emphasize engagement in preserving and, in some cases, rebuilding it through sustainable local and global business approaches. In addition, the social component is a detailed feature of company proposals that is linked to the mission, responsibility, and company impact in a broader way than their narrow business context.

> The values advanced by B Lab and B-corporations are ones that are important to us and also to our stakeholders. By becoming a B-corporation, we are better able to broadcast our commitment to those values and hopefully set an example of social and environmental responsibility in our community. (documents from Firm N. 26)

> Our company, through the lines of work on issues of urban agriculture, environmental education and sustainable community development, seeks to create better cities and responsible consumers. (documents from Firm N. 35)

10.4.2 Practices Linked to Value Propositions

The principles, standards, ethics, and ideals of B-corporations conveyed in the value propositions guide the actions of both the organizations and their people.

These companies strive to have a positive impact on society by considering how to benefit their business context. Constructive change projects begin with an examination of how firms can contribute to challenging the status quo of traditional business and social practices; such evaluation is synthesized in the so-called 'change we seek', which is a sort of call to action. Three main practices are dominant, and we named as follows: (1) creating collective sense, (2) supporting actors, and (3) improving contexts.

A first practice expressed by B-corporations as of critical concern within their value propositions pertains to creating a collective sense of change. B-corporations strive to build a shared meaning of what a better company should be with respect to customers', stakeholders', and other actors' needs by aligning their beliefs with economic, social, and environmental values as well as by providing clear and relevant evidence of the benefits of sustaining responsible social and business practices. The following quotes illustrate how companies attempt to elicit conscious support for responsible behaviour from their customers:

> Our proposal is to propagate the responsible use of solar energy, pioneer conscientious business practices and create holistic wealth for ourselves and our community offline. (documents from Firm N. 6).

> The company challenges professionals from different fields to work together and create a sustainable project plan. It accomplishes this collaboration by holding a Green Design Charrette—a facilitated, multi-day design event that challenges diverse groups of professionals and community members to collaborate on developing the most sustainable plan/design for a given project offline. (documents from Firm N. 14)

B-corporations seek to unlock resource potential by actively *supporting* actors in their responsible change practices. In doing so, they provide actors with tutoring, support to improve the environmental and social aspects of activities, and, in some cases, extending the same services to customers and communities of B-corporation clients. In a similar way, many companies establish and diffuse sustainability standards in their industry and supply chain.

> We became a Certified B-corporation: to inspire other companies to adopt socially responsible practices, and to be inspired by others; to allow our stakeholders to know that (the company) is a socially responsible company; and to learn how (the company) can be more socially responsible. (documents from Firm N 22).

> The company provides services and tools to support the design, construction, and operation of buildings that are responsible, enduring, and healthy. With its collective expertise and open exchange, it helps clients integrate sustainable practices that benefit their business, their community and the environment. (documents from Firm N. 33).

> The company and its partners have developed innovative ways to encourage members to live a greener lifestyle. This includes 'Green Your' challenges that leverage the powerful concept of 'gaming for good' to engage members online in a fun, educational environment while influencing eco-friendly behaviors offline. (documents from Firm N. 13).

Finally, B-corporations work to *improve* the context they share in order to remain close to their economic, sustainability, and social goals and values. The extent to which B-corporations articulate their efforts relies on their ability to signal to customers, shareholders, investors, and other actors what to expect from the companies' activities. They clearly provide their way to correspond with how they act. In this sense, B-corporations clearly and systematically communicate their social and environmental outcomes and performances in order to 'walk their talk' about being a better company. They act with the understanding that each actor is dependent on upon another, and they are thus responsible for each other and for future generations.

> Our business is to protect and improve the quality of life in the communities we serve. Our motto is Serve Others. We believe that business can and should be a powerful force for positive social change. (documents from Firm N. 4)

> The company seeks to unlock the potential of urban small businesses to create value in the marketplace; produce and retain jobs; improve standards of living; and produce sustained economic vitality in their communities. (documents from Firm N. 35)

10.4.3 Value Propositions and B-corporations Ecosystems

B-Lab and B-corporations enact shared institutional logics between all actors by disseminating common norms and values on which they strategize and act.

> We became a B-corporations because we saw real value in embracing a 3rd party assessment focused on benchmarking socially aware businesses. Through the assessment, we can take a step back and analyze our business according to a balanced scorecard, taking all ecosystem stakeholders into account. Through the community of B-corporations we can expand our ecosystem to include like-minded businesses, improving our ability to deliver on our Social Business vision. (documents from Firm N. 12)

Within the analysed context, the ecosystem is not an abstract concept but represents a reality to be developed and preserved, with local and global implications involving different actors. B-corporations' value propositions tackle multiple value-creation contexts, not just for users or main stakeholders, but for the whole ecosystem of actors.

> We want to generate a positive impact inside the ecosystem in which we operate by improving the quality of community life. (documents from Firm N. 26)

> We aim to create an [...] ecosystem filled with companies that matter to the world, because the world matters to them. (documents from Firm N. 32)

B-corporations envision their value propositions as a unified goal that considers values as the building block for creating an ecosystem of value practices among all actors. The Impact Report collected from companies includes the following quote:

> Transforming How We Make Money While Making a Difference. We need to work on to create the system/framework for B-corporations [. . .] We strongly believe that this is the way to go, we need to change the way we are doing all kind of businesses. (documents from Firm N. 23)

The value propositions highlight responsibility in the business as well as the social context so that goals are not only profit-oriented, but also include benefits for everyone in the service ecosystem. Companies strive to make the values of responsibility achievable by proposing their idea of what a better company should be, and working to implement this idea in practice within the ecosystem.

> Our aim is to grow a fluid and mobilised global ecosystem where any enterprising idea can enter at a local level and immediately scale sustainable impact. (Documents from Firm N. 31)

> We strive to create products that protect health but do so at no net cost to our planet or our quality of life. They aim to enrich the lives of their users, their retailers, their stakeholders and their neighbours. This is our Ecosystem and the basis of their triple bottom line. (Documents from Firm N. 10)

B-corporations are not simply firms that voluntarily meet higher standards of transparency, accountability and performance; they also work to build a collective voice through the power of the unifying values, rules, and practices. By breaking free from traditional business in terms of actions, language, norms, and performances, they are actively dedicated to creating shared value for the ecosystem.

> The company is driven by sustainable prosperity and by cultural, social and ecosystem regeneration. In a transition toward green and social economies, the company creates community and sustainable value with its clients, suppliers, competitors and strategic partners. (documents from Firm N. 2).

> Leading the business revolution is only one step to creating true, positive change—we must, as all companies must, uphold a values system that integrates social, economic and environmental wellbeing. (documents from Firm N. 13).

10.5 Discussion

This work aimed to understand the contributions of actors' values, aims, and practices towards a service ecosystem. The authors focused on the role of value propositions by investigating the community of B-corporations. In order to answer to the research question (What is the role of value propositions in shaping a service ecosystem?), this work built on performativity research (Kjellberg and Helgesson 2006) and analysed the performative role of value propositions as a bridge between actors' values, aims, and practices in shaping a service ecosystem.

B-corporations' value propositions are built on the companies' values, with specific reference to social and environmental issues. They are not only committed to providing add-on benefits for customers in order to achieve strategic advantage, but also involve the diffusion of values, norms, and practices that surpass the offering of new or better products/services. Value propositions enable compelling and self-powered experiences of what a good business should be, which are capable of regenerating the ecosystem. Regeneration is a key issue in the value propositions of B-corporations and this feature leads to favour a responsible use of resources towards the creation of value for all stakeholders. This consideration is in line with the theoretical proposal by Greer et al. (2016), as value is addressed through value proposition to external customers and all the other stakeholders.

By espousing and performing a new logic of conducting business, the B-corporations' value propositions assume a performative role. Following Kjellberg and Helgesson's (2006) notion of *performativity*, this work suggests viewing the role of value propositions in shaping the ecosystem as framed by a multiplicity of actors who share collective values and perform common practices (Hagberg and Kjellberg 2010). Main contribution of this work is that it addresses how ideas—which are the pre-constitute aspects of the ecosystem such as values, norms and meaning—participate in shaping reality (i.e. the ecosystem) through the translation of such ideas into practices. In understanding 'how actors influence reality' (Storbacka et al. 2012, p. 69) theoretical constructs are not simply a guide to describe reality, but actively contribute to the construction of reality by shaping actual practices.

A model depicting the performative role of value propositions in shaping service ecosystem is shown in Fig. 10.2. It consists of two levels:

1. An idea level, where values shape value propositions;
2. A practice level, where value propositions shape practices.

Concerning the idea level, the findings show that B-corporations contribute to shaping the service ecosystem through their ideas and principles regarding a better way of doing business in terms of environmental, social, economic health, and responsibility dimensions. Such collective beliefs and values shape value propositions, which bring multiple benefits: economic performance enables the achievement of environmental and societal outcomes. Companies offer values, not simply

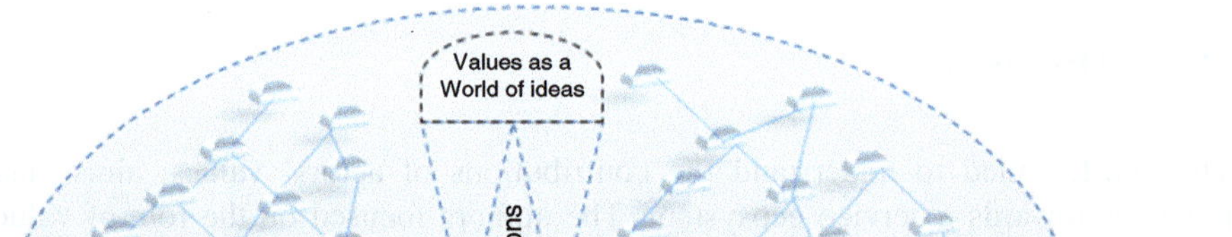

Fig. 10.2 The performative role of value propositions. Source: Authors' elaboration

goods and services, and seek a change within the service ecosystem in terms of rules, norms, and practices for all actors.

B-corporations' value propositions address a context in which every actor is connected to another through collective values. In accordance with (Payne and Frow 2014a) and Storbacka and Nenonen (2011), value propositions are formed by transforming the logic of individual value creation—for the customers and/or for the companies—into a more integrated viewpoint that emphasizes all the actors involved in the ecosystem (Gummesson and Mele 2010). Values are the glue that links together companies and with other actors (Edvardsson and Enquist 2009), whereas the creation of value is set in wider actors to actors or the ecosystem context (Vargo and Lusch 2016). Value propositions are a tool for joining actors (Frow and Payne 2011; Lusch and Webster 2011; Payne and Frow 2014b) in collective sense-making activities aimed at framing a shared ecosystem view (Storbacka and Nenonen 2011). By disseminating the common norms and values on which they strategize and act, B-Lab and B-corporations enact shared institutional logics that further connect stakeholders and all other actors, and enable mutual value creation (Vargo and Lusch 2012). Moreover, collective norms and resources integration are crucial in favouring new value propositions and the reshaping of ecosystems (Frow et al. 2016; Greer et al. 2016). In this perspective the reciprocity of value propositions (Ballantyne et al. 2011) mirrors the common interests of all stakeholders and the interest in achieving goals aimed by various actors. These goals are represented by the 'doing good', 'acting for good' and 'making a difference' in B-corporations principles.

The second level focuses on practices, and provides understanding of how firms act to face current economic and societal challenges. B-corporations' value propositions address not only values, but also the practices involved in implementing those values within the community. More than a promise of future benefits, they are a declaration of actions—i.e., a performative utterance. More specifically, B-corporations make ideas concrete by performing actions in order to impact on the way they do business. The practices we named—*creating a collective sense*—depict these efforts. B-corporations do not simply adopt responsible behaviours, but spread their vision and act to inspire other actors to adopt common ways of sustaining and translating values into the realm of practices. The vision of what it means to be a better company is supported by the premise that all actors are mutually dependent on one another. Practices are led by a business vision that relies on the actions of interconnected actors who depend on one another for their mutual effectiveness and wellbeing. This can be seen in the B-corporations' practices of *supporting actors* and practices of *improving their context.* Through their value propositions, B-corporations attempt to change the practices of the service ecosystem and to align the respective interests of all actors. A community arises, whereby actors actively participate in the promotion of shared values and meanings in a collective context and support all other actors in performing in accordance with them. This view is line with the study by Payne and Frow (2014a) addressing the service ecosystem as the context hosting the emerging of value propositions and their effect on achieving wellbeing.

As Fig. 10.2 shows, circular relations between the worlds of ideas and the worlds of practices occurs by means of value propositions. Values inform value propositions, which can be seen as narratives or discourses of value potential (Vargo and Lusch 2016) that affect actors' practices in shaping the service ecosystem. Value propositions are discourses that connect the worlds of idea—what a better business can be—to the word of practices—what a better ecosystem is.

In sum, this work promotes a conceptualization of value propositions that is neither firm- nor customer-centric, but that revolves around values and practices that a multiplicity of actors espouses and performs in their value processes. Such value proposition links ideas about the service ecosystem to the real service ecosystem.

10.6 Main Implications

This research highlights the *performative* role of value propositions, which lies in the ability of values and practices to shape a service ecosystem. In this view, value propositions act on the worlds of values and practices concerning what an ecosystem can be and what it effectively is. In line with Kjellberg and Helgesson (2006), this study illustrates how envisioning a world of ideas and developing a world of practices can affect the wider service ecosystem.

The results have implications for both managers and scholars. For managers, the need to develop value propositions by adopting a service ecosystem perspective is addressed. If a company's focus is on the service ecosystem as a whole and not simply an individual or company's perspective, then value propositions will be necessarily intermeshed with the practices and values of the whole ecosystem. This means that all actors should be seen as co-responsible for others, in the same way that they co-create value for themselves. Managers can contribute to the alignment of value, interests, and practices of all actors within the service ecosystem. They should be aware that companies' strategies are tools that shape the service ecosystem by seeking changes in the ideas and actions of a number of actors to achieve shared values and practices that support the viability of the ecosystem as a whole.

The service ecosystem perspective can be employed to understand the interaction between actors in a multi-level context where the concepts of survival and the health of the ecosystem are intrinsically linked to an actor-centred perspective of value. This perspective takes into account the social norms and institutional structures that support ecosystem viability. Ideas and practices can be seen as means by which to grasp how companies support and perform their strategy within the service ecosystem; additionally, the roles of all stakeholders in contributing to change practices, norms and the ideas of value should be considered by managers in reshaping the content of value propositions.

In this view, managers could work on developing value propositions not as supplier promises to customer or stakeholders, but as an agreement among multiple actors to shape the service ecosystem. The focus should be on the matching process between values and practices.

For scholars, the *performativity* view contributes to the extant research by expanding the conceptualization of value propositions (Payne and Frow 2014b; Vargo and Lusch 2016). This work offers a fresh perspective on the role of value propositions in disseminating values that influence actors' practices in shaping service ecosystems (Payne and Frow 2014b; Vargo and Lusch 2016). Value propositions are not simply promises of reciprocal or mutual benefits (Ballantyne et al. 2011); they assume a performative role in actors' aims and actions. Thus, performativity of value propositions through linking ideas and practices together gives substance to the strategy of companies within the service ecosystem, and contributes to ecosystem viability. Value propositions represents the way leading to a more sustainable service ecosystem, due to the involvement of actors in integrating resources, the institutional change in supporting ecosystem viability (Meynhardt et al. 2016), and the sharing of practices by multiple actors in a shared context (Wieland et al. 2016).

However, studies on this topic are still in their infancy and the extant and recent literature is further calling for investigations on how actors can be part of a service ecosystem (Frow et al. 2016) and on the need to consider value co-creation in a society-based perspective (Meynhardt et al. 2016). Some questions that remain to be addressed in future research are: What are the antecedents and mechanisms according to which companies' values and practices affect customers' and other actors' values and practices? How can different actors' values and practices be aligned? What implications do the interrelationships between different ideas and practices have for a unified ecosystem's value proposition? What effect do changes in one context exert on another from a multi-actor perspective? How may the performativity of value propositions change according to different contexts and under several conditions? Are there moderating factors that are able to influence the performativity of value propositions?

References

Austin, J. L. (1962). *How to do things with words*. Oxford: Oxford University Press.

Ballantyne, D., Frow, P., Varey, R. J., & Payne, A. (2011). Value propositions as communication practice: Taking a wider view. *Industrial Marketing Management, 40*(2), 202–210.

Barile, S., & Polese, F. (2010). Smart service systems and viable service systems: Applying systems theory to service science. *Service Science, 2*(1-2), 21–40.

Barile, S., Lusch, R., Reynoso, J., Saviano, M., & Spohrer, J. (2016). Systems, networks, and ecosystems in service research. *Journal of Service Management, 27*(4), 652–674.

Battilana, J., & Lee, M. (2014). Advancing research on hybrid organizing – Insights from the study of social enterprises. *The Academy of Management Annals, 8*(1), 397–441.

Butler, J. (2011). *Bodies that matter: On the discursive limits of sex*. London: Taylor & Francis.

Callon, M. (1998). Introduction: The embeddedness of economic markets in economics. In M. Callon (Ed.), *The laws of the markets*. Oxford: Blackwell.

Cavell, S. (2002). *Must we mean what we say?* (updated edition). Cambridge: Cambridge University Press.

Chabowski, B. R., Mena, J. A., & Gonzalez-Padron, T. L. (2011). The structure of sustainability research in marketing, 1958–2008: A basis for future research opportunities. *Journal of the Academy of Marketing Science, 39*(1), 55–70.

Crittenden, V. L., Crittenden, W. F., Ferrell, L. K., Ferrell, O. C., & Pinney, C. (2011). Market oriented sustainability: A conceptual framework and propositions. *Journal of the Academy of Marketing Science, 39*(1), 71–85.

Diaz Ruiz, C. A. (2012). Theories of markets: Insights from marketing and the sociology of markets. *The Marketing Review, 12*(1), 61–77.

Drucker, P. F. (1954). *The practice of management*. Cambridge: Routledge.

Dubois, A., & Gadde, L. E. (2002). Systematic combining: an abductive approach to case research. *Journal of Business Research, 55*(7), 553–560.

Edvardsson, B., & Enquist, B. (2009). *Values-based service for sustainable business, lessons from IKEA*. London: Routledge.

Edvardsson, B., Tronvoll, B., & Gruber, T. (2011). Expanding understanding of service exchange and value co-creation: A social construction approach. *Journal of the Academy of Marketing, 39*, 327–339.

Edvardsson, B., Enquist, B., & Sebhatu, S. P. (2013). The role of shared values in a sustainable service business. In R. P. Fisk, R. Russell-Bennett, & L. C. Harris (Eds.), *Serving customers* (pp. 229–248). Prahran Vic: Tilde University Press.

Edvardsson, B., Kowalkowski, C., Strandvik, T., & Voima, P. (2014). Negative critical waves in business relationships: An extension of the critical incident perspective. *Journal of Business and Industrial Marketing, 29*(4), 284–294.

Enquist, B., Johnsson, M., & Skålén, P. (2006). Adoption of corporate social responsibility – Incorporating a stakeholder perspective. *Qualitative Research in Accounting and Management, 3*(3), 188–207.

Frow, P., & Payne, A. (2011). A stakeholder perspective of the value proposition concept. *European Journal of Marketing, 45*(1/2), 223–240.

Frow, P., McColl-Kennedy, J. R., Hilton, T., Davidson, A., Payne, A., & Brozovic, D. (2014). Value propositions A service ecosystems perspective. *Marketing Theory, 14*(3), 327–351.

Frow, P., McColl-Kennedy, J. R., & Payne, A. (2016). Co-creation practices: Their role in shaping a health care ecosystem. *Industrial Marketing Management, 56*, 24–39.

Greer, C. R., Lusch, R. F., & Vargo, S. L. (2016). A service perspective. *Organizational Dynamics, 1*(45), 28–38.

Gummesson, E. (2005). Qualitative research in marketing: Road-map for a wilderness of complexity and unpredictability. *European Journal of Marketing, 39*(3/4), 309–327.

Gummesson, E. (2008). Customer centricity: Reality or a wild goose chase? *European Business Review, 20*(4), 315–330.

Gummesson, E., & Mele, C. (2010). Marketing as value co-creation through network interaction and resource integration. *Journal of Business Market Management, 4*(4), 181–198.

Hagberg, J., & Kjellberg, H. (2010). Who performs marketing? Dimensions of agential variation in market practice. *Industrial Marketing Management, 39*, 1028–1037.

Hall, S. (2000). Who needs identity? In P. du Gay, J. Evans, & P. Redman (Eds.), *Identity: A reader*. London: Sage Publications.

Hsieh, H.-F., & Shannon, S. E. (2005). Three approaches to qualitative content analysis. *Qualitative Health Research, 15*(9), 1277–1288.

Iansiti, M., & Levien, R. (2004). Strategy as ecology. *Harvard Business Review, 82*(3), 68–78.

Jackson, S. (2004). *Professing performance: Theatre in the academy from philology to performativity*. Cambridge: Cambridge University Press.

Kjellberg, H., & Helgesson, C. F. (2006). Multiple versions of markets: Multiplicity and performativity in market practice. *Industrial Marketing Management, 35*(7), 839–855.

Kjellberg, H., & Helgesson, C. F. (2007). On the nature of markets and their practices. *Marketing Theory, 7*(2), 137–162.

Korkman, O., Storbacka, K., & Harald, B. (2010). Practices as markets: Value co-creation in e-invoicing. *Australasian Marketing Journal, 18*(4), 236–247.

Koskela-Huotari, K., Edvardsson, B., Jonas, J. M., Sörhammar, D., & Witell, L. (2016). Innovation in service ecosystems—Breaking, making, and maintaining institutionalized rules of resource integration. *Journal of Business Research, 69*(8), 2964–2971.

Kowalkowski, C. A. (2011). Dynamics of value propositions: Insights from service dominant logic. *European Journal of Marketing, 45*(1), 277–294.

Kozinets, R. V. (2002). The field behind the screen: Using netnography for marketing research in online communities. *Journal of Marketing Research, 39*(1), 61–72.

Latour, B. (2005). *Reassembling the social: An introduction to actor-network-theory*. Oxford: Oxford University Press.

Law, J., & Urry, J. (2004). Enacting the social. *Economy and Society, 33*(3), 390–410.

Lincoln, Y. S., & Guba, E. G. (1985). *Naturalistic inquiry* (Vol. 75). London: Sage.

Loxley, J. (2006). *Performativity*. Abingdon: Routledge.

Lusch, R. F., & Vargo, S. L. (2014). *An introduction to service-dominant logic*. Cambridge: Cambridge University Press.

Lusch, R. F., & Webster, F. E. (2011). A stakeholder-unifying, cocreation philosophy for marketing. *Journal of Macromarketing, 31*(2), 129–134.

Lusch, R. F., Vargo, S. L., & Gustafsson, A. (2016). Fostering a trans-disciplinary perspectives of service ecosystems. *Journal of Business Research, 69*(8), 2957–2963.

Meynhardt, T., Chandler, J. D., & Strathoff, P. (2016). Systemic principles of value co-creation: Synergetics of value and service ecosystems. *Journal of Business Research, 69*(8), 2981–2989.

Nordgren, L. (2008). The performativity of the service management discourse: Value creating customers in health care. *Journal of Health Organization and Management, 22*(5), 510–528.

Payne, A., & Frow, P. (2014a). Developing superior value propositions: A strategic marketing imperative. *Journal of Service Management, 25*(2), 213–227.

Payne, A., & Frow, P. (2014b). Deconstructing the value proposition of an innovation exemplar. *European Journal of Marketing, 48*(1/2), 237–270.

Peterson, M. (2015). Social enterprise for poverty alleviation in an era of sector convergence. *Journal of Ethics and Entrepreneurship, 5*(1), 5–25.

Piekkari, R., Plakoyiannaki, E., & Welch, C. (2010). Good' case research in industrial marketing: Insights from research practice. *Industrial Marketing Management, 39*(1), 109–117.

Russo Spena, T., Mele, C., & Nuutinen, M. (2017). *Innovating in practice*. Cham: Springer International Publishing.

Sabeti, H. (2011). The for-benefit enterprise. *Harvard Business Review, 89*(11), 98–104.

Skålén, P., Gummerus, J., von Koskull, C., & Magnusson, P. (2015). Exploring value propositions and service innovation: A service-dominant logic study. *Journal of the Academy of Marketing Science, 43*(2), 137–158.

Storbacka, K., & Nenonen, S. (2011). Scripting markets: From value propositions to market propositions. *Industrial Marketing Management, 40*, 255–266.

Storbacka, K., Frow, P., Nenonen, S. M., & Payne, A. (2012). Designing business models for value co-creation. *Review of Marketing Research, 9*, 51–78.

Taillard, M., Peters, L. D., Pels, J., & Mele, C. (2016). The role of shared intentions in the emergence of service ecosystems. *Journal of Business Research, 69*(8), 2972–2980.

Vargo, S. L., & Akaka, M. A. (2012). Value cocreation and service systems (re) formation: A service ecosystems view. *Service Science, 4*(3), 207–217.

Vargo, S. L., & Lusch, R. F. (2011). It's all B2B and beyond...: Toward a system perspective on the market. *Industrial Marketing Management, 40*, 181–187.

Vargo, S. L., & Lusch, R. F. (2012). *Service-dominant logic: A forward look*. Presentation at The Tekes-serve annual seminar, Helsinki 26-04-2012.

Vargo, S. L., & Lusch, R. F. (2014). Inversions of service-dominant logic. *Marketing Theory, 14*(3), 239–248.
Vargo, S. L., & Lusch, R. F. (2016). Institutions and axioms: an extension and update of service-dominant logic. *Journal of the Academy of Marketing Science, 44*(1), 5–23.
Vargo, S. L., Wieland, H., & Akaka, M. A. (2015). Innovation through institutionalization: A service ecosystems perspective. *Industrial Marketing Management, 44*, 63–72.
Weber, R. P. (1990). *Basic content analysis* (2nd ed.). Newbury Park: Sage.
Wieland, H., Polese, F., Vargo, S. L., & Lusch, R. F. (2012). Toward a service (eco) systems perspective on value creation. *International Journal of Service Science, Management, Engineering, and Technology, 3*(3), 12–25.
Wieland, H., Koskela-Huotari, K., & Vargo, S. L. (2016). Extending actor participation in value creation: an institutional view. *Journal of Strategic Marketing, 24*(3-4), 210–226.

Chapter 11
A Journey Through Possible Views of Relational Logic

Marco Pellicano, Maria V. Ciasullo, Orlando Troisi, and Gian Luca Casali

Abstract This chapter discusses how the relational logic is one of the most interesting and profitable management paradigms to understand firms' behaviour and to foster value co-creation processes.

The relational view and the resulting relational inter-subjective dynamics are critical success factors for corporate governance capable to mobilise different involved actors' resources. However, the importance of a system perspective of relational dynamics is only recently discussed in the current doctrinal debate. Therefore, our goal is to contribute in bridging this gap by reinterpreting, in a system key, the issue of relations. Methodologically, the authors have carried out a review taking a system vision, in order to conceptualize an original interpretative model. The interpretative model that we call "Enterprise Relational Vision" (ERV) is based on the following assumptions (pillars): Relationality and autopoiesis, Dissemination and definition, Sense and cohesion, Training and decision formulation, Co-creation and regeneration, Resources and competitiveness, Leadership and viability.

Keywords Relational logic • System view • Stakeholders • Value creation

11.1 Introduction

During the 1990s of the last century, a multidisciplinary group of researchers belonging to the University of Salerno and adhering to system thinking promoted the development of an "observational perspective" aimed at outlining an "interpretative model" of entrepreneurial social phenomena, able to better understand the complexity of their governance and to give the right emphasis to their inner "relationships". The appliance of the so-called relational view to the observation of complex socioeconomic phenomena has spread over the last decades. In

M. Pellicano (✉) • M.V. Ciasullo • O. Troisi
Department of Business Science – Management and Innovation Systems, University of Salerno, Fisciano, Italy
e-mail: pellicano@unisa.it

G.L. Casali
School of Management, Queensland University of Technology, Brisbane City, QLD, Australia

S. Barile et al. (eds.), *Social Dynamics in a Systems Perspective*, New Economic Windows, DOI 10.1007/978-3-319-61967-5_11

particular, it has been applied to some specific subject areas such as *the relational theory of society* (Donati 2004), considered the emerging paradigm of modern sociology, or *the relational theory of happiness* (Bruni and Zamagni 2004). In business studies, the *relational logic* represents one of the most interesting and profitable management paradigms; thus, it represents a fascinating research perspective, capable to foster the understanding of firm behaviour. In managerial domain, marketing studies have some sort of "primacy" in accepting the relational perspective (Berry 1983), especially when dealing with the interpretation of business-to-business (Håkansson 1982; Håkansson and Snehota 1995) and business-to-consumer (Grönroos 1983; Grönroos and Gummesson 1985) markets. In the current competitive scenarios, the appliance of the relational view represents a critical success factor for managerial decision-making, being able to enhance critical aspects such as values, culture, identity, sharing, and cohesion. In fact, both doctrine and managerial practice consider the relations between companies as strategically important (Håkansson 1982). This importance derives from their significant contribution to the joint creation of value for market (Normann and Ramírez 1994), the creation of economic capital (Vicari et al. 2000), the achievement and sustainability of competitive advantage (Dyer and Singh 1998; Nahapaiet and Goshal 1998), as well as the definition of innovative approach to business development. Therefore, the resulting relational inter-subjective dynamics can be read as relevant socio-economic phenomena, able to mobilise in a synergistic way the resources of different involved actors' in order to foster mutual/shared value creation processes. The relational vision is also in line with the theories recently emerged in Service Research, such as Service Dominant Logic (Vargo and Lusch 2004) and Service Science (Maglio and Spohrer 2008), that focus on the importance of service, seen as the basis of the exchanges among the actors participating in co-creation processes.

To offer a better interpretation of complex social phenomena, a system perspective of relational dynamics is needed in the analysis of relational enterprises. However, this perspective has recently interested the current academic debate. Therefore, this work is aimed at entering in this research stream, providing an "interpretative model" able to read the issue of relations according to a system approach.

The authors defined this interpretative model "Enterprise Relational Vision" (ERV). It represents an original observational perspective of entrepreneurial phenomena and puts emphasis on the importance of intangible resources linked to skills and competence, those contextual relations aimed at developing the company's and its knowledge capital, the responsible involvement of stakeholders in value co-creation processes and the trust in the relationship.

In terms of methodologic approach and to conceptualize an original interpretative model, drawing on a system vision, the authors have carried out a review of main academic contributions related to different scientific domains useful to their conceptualizations. The followed approach has attempted to integrate the recent system theory, including the Parsons and Luhmann's "Social System", the Beer's

"Viable System", the Capra and Luisi's "Systems of Life" and the Maglio and Spohrer's "Service Science".

This chapter discusses the possible views related to the relational logic, which, despite its multifaceted analysis enhances the relationship between two parties: the *Ego*, as the decision maker that assumes a central observational position in the system, and, the *Alter*, as a part of the interpersonal relationship network of the *Ego*. Each of the following views is discussed as resulting from multiple theories that, despite they were created in different times and, perhaps, with different aims, could be combined to better explain all possible features of the relational logic.

Those views are:

- *Relationality and autopoiesis*: the relational nature of communication processes constitutes and supports firms, which are considered an autopoietic social system.
- *Dissemination and definition*: the reticular structure and its emergent system arise from the *Ego* (decision maker) and take place in the specific enterprise context, which represents a "relative portion" of the general environment (eco-system). Drawing on the dual perspective of structure and system, company is built upon the constructive observation of an *Ego*.
- *Sense and cohesion*: company is considered a system with a symbolic sense, represented by value co-creation. The sense of belonging stems from a complex mix of rational and emotional factors.
- *Decision formation and formulation*: in terms of complexity of problems, the decision-making seems to be quite complex where the decision power tends to be shared and spread among different roles, that are all *Altera* (entities in relation with the *Ego*).
- *Co-creation and regeneration*: value co-creation, internal to the relational network that represents the organizational pattern of a system, allows the autopoietic regeneration of resources at the roots of its viability.
- *Resources and competitiveness*: the competitiveness of a social system is strictly related to the ability in gaining the needed resources establishing collaborative relationships, i.e. relationships which frame them.
- *Leadership and viability*: the Governing Subject guarantees the relational harmony, which is characterized by the environmental dynamism. Therefore, its continuous guidance is fundamental in facilitating communication processes, which ultimate goal is the viability of the social system.

The following sections delve on a wider exposition of the above-mentioned considerations.

11.2 The Origin of System Thinking

Reading the firm according to a system perspective cannot ignore the "System Thinking" and the related "System Theory" as well as their most recent advancements.

In particular, the system thinking originates from Greek philosophy and in particular from Aristotle and the early Pythagoreans conceptualization, whose reflections focused on the dichotomy between the shape and the substance of things. These concepts root on the difference among components (substance) and relationships (shape).

The General System Theory was born in the middle of the last century thanks to some scholars coming from different scientific domains (mathematical, biological and sociological) and striving for confrontation, the progressive enrichment of the knowledge about common and shared meaning of system and combining different perspectives of analysis due to their different disciplinary backgrounds. This initiative comes from the concept of system, coined by Von Bertalanffy (1968) and defined as "a set of elements interacting with each other". This definition offers a reinterpretation and an adaptation to the investigation and the analysis which can be conducted in each specific discipline. Another element fundamental to the understanding of the general system theory can be identified in the reasons of its emergence. The system approach has been defined to counteract the tendency of natural processes toward entropy that is toward the disruptive disorder (Pardi 1998); recently aggravated by the emergence of the complexity that deeply affects the rationality of decision-making processes. Reading a phenomenon according to a system perspective leads the seeking for an order that can be applied to the analysis of any organization and can foster the achievement of a functional equilibrium over time (Pels et al. 2014; Mele et al. 2010).

Based on the most recent advancement of system thinking, some alternative types of "systems of life" (Capra and Luisi 2014), antithetical if compared to "mechanical systems", can be identified:

- Biological systems (which we define "living");
- Social systems (which we define "life").

Due to the increasing complexity, both of these can be considered "cognitive" systems, when able to self-organize (Ashby 1962), or "cybernetic" systems, when equipped with their own self-regulating mechanisms (Wiener 1948; Beer 1989) and, in any case, able to interact and co-evolve with the broader "ecosystems" in which they are embedded. Consequently, different dimensions of life can be observed; thus, this general paradigm shift lies on the different metaphors that illustrate the scheme of life: "from machine to network".

11.3 Relationality and Autopoiesis

The term Autopoiesis refers to the ability of internal regeneration, which characterizes all the systems of life, notwithstanding their specificity.

In social systems, the autopoiesis is based on "communication"; thus, the living networks of human society are considered as "communication networks" (Luhmann 1984), while biological systems exchange molecules, social systems exchange ideas, information, knowledge, skills (i.e. intangible resources) in their communicative relations networks (Capra and Luisi 2014).

If biological systems produce and nurture a material boundary (cell membrane), social systems produce and nurture a non-material boundary (cultural, worth, utilitarian) which constrains the behaviour of its members (Luhmann 1984). Therefore, the autopoiesis refers to life maintenance; thus, a (biological or social) system stays alive as long as it is able to regenerate itself.

In our view and in line with Service Dominant Logic, "services" are analogous to the molecules of biological organisms (Vargo and Lusch 2004), i.e. resources integrated and delivered through relationships.

Finally, the cognitive ability (cognition) represents the system ability to interact with the environment profitably, not necessarily "smart", as well as the ability to co-develop harmoniously (Capra and Luisi 2014).

In sum, the (biological and social) systems of life are characterized by the following substantial similarities:

- The importing of several forms of general energy from the environment (e.g. air, light, climate in its meteorological, social and economic meanings);
- The gaining of specific nutritional resources from the environment (alimentary or economic, material or immaterial);
- The elaboration and transformation of resources through "metabolic autopoietic processes" (chemical or communicative) self-organized into "networks", in which basic element (molecules or services) are exchanged through "relationships";
- The creation of new energy (biological and socioeconomic) which, in turn, generates "viability" and, ultimately, supports "life".

The firm completely assumes its natural dimension of social system just under a specific condition, that is the emergence of reticular interconnections among its components that can be considered "relational", being oriented to the collaboration (Pellicano 1994).

According to our view, Luhmann's social autopoiesis is completely realized, if "communication" takes the features of a "circular dialogue", which is empathic and, consequently, relational (Bateson 1979; Pellicano 1992). In other words, following the logic at the roots of molecular exchanges typical of biological autopoietic processes, the development of a relational network (enabled for resource exchange) represents the core element of those mechanisms that led to the on-going regeneration of firms' viability. Therefore, the disposition toward a

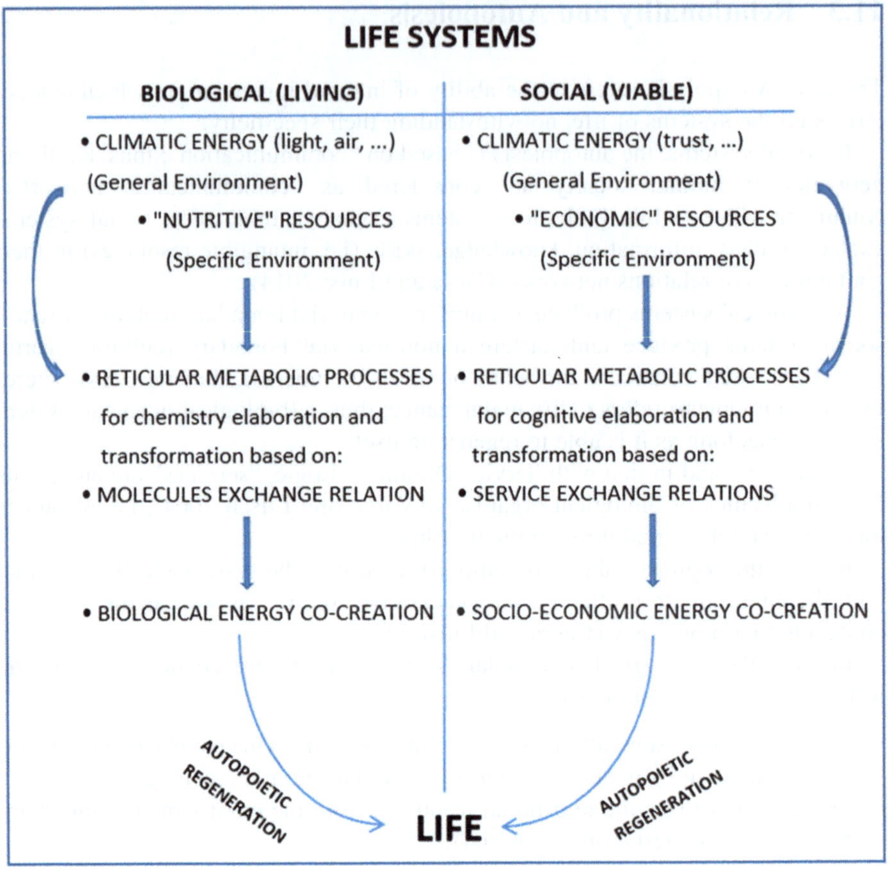

Fig. 11.1 A comparison of life systems

respectful and constructive dialogue, supported by strong internalized "values and ethical principles", represents a clear expression of the "wisdom" of those viable firms that aim at surviving (Pellicano 2005) (Fig. 11.1).

For the purpose of our study, we consider essential to achieve a conceptualization of the nature and the meaning that we attribute to the term "relation". Primarily, it should be underlined that relations are included in the broader category of rapports, which, according to their structural sense, are defined in terms of connections, while according to their systemic meaning, are defined as interactions.

Contention and exchange are the two the main forms that rapports can assume. In the first case, we refer to the category of antagonistic rapports based on a win-lose logic, which, following a degenerative path, evolve in destructive rapports based on lose-lose logic. In the second case, rapports can be occasional (or powered by an exchange based on a mere economic convenience), symbiotic (when the dependence of a subject on another one characterizes the exchange) or explanatory. We define a rapport "relation" just in the latter case, because it roots on a "synergistic" exchange that emphasize the collaborative nature and the disposition towards temporally long-lasting rapports. Specifically, relations arise from the structural connection among two or more parties, which, at a system level, evolve towards interactions, characterized by a common and collaborative path that is resonance (Beer 1989).

Drawing on a cybernetics perspective, the elements on which a system is built up are combined in a mutual interaction, whereby the action of an element on another implies a response (feedback or reply) of the latter towards the former. Therefore, it possible to assume that these two elements are linked by a "feedback belt". The belt binding the element A to the element B is defined "positive" (relational), because a variation in the value assigned to A makes a similar change to the value assigned to B (Wiener 1948).

Focusing on the meaning that an integration logic gives to exchanges, some issues related to the subjects involved in the entrepreneurial action arise. According to the relational logic, an exchange generally involves two types of actors: (1) the *Ego*, which is the decision-maker that assumes a central observational position within the system; (2) the *Alter*, which is, for example, an interpersonal relationship network of the *Ego*.

Which is the object of active exchanges between these parties?

The answer is in the resources commonly understood as involved in the process of value creation (Fig. 11.2).

The system offers to each actor a value output generated by relational interactions, which represents an incoming resource for the subject, who holds and potentially bears resources. Consequently, in an input/output exchange, value and resources represent two sides of the same coin.

By resorting to a metaphor, the actors can be considered as islands that share the same sea, which represents the common environment. It is evident that islands should not be isolated, so they build bridges, which are the more or less stable links among them. The bridge represents the structural connection that should be

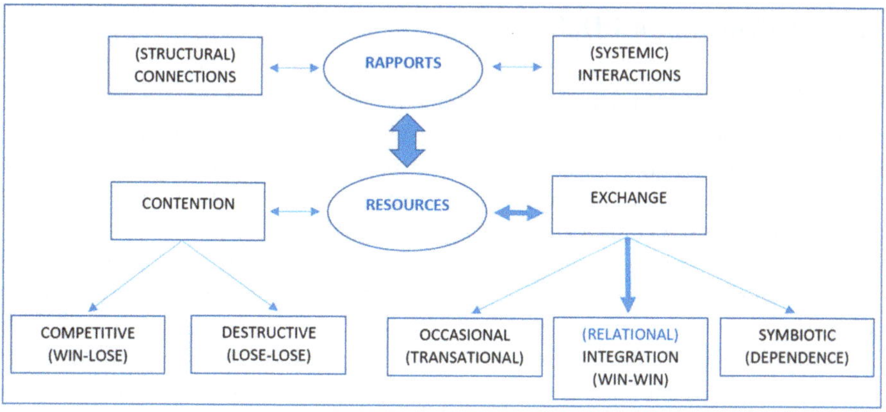

Fig. 11.2 From connections to relations

animated by those interactions aimed at fostering resource exchange and, consequently, value exchange. Therefore, interactions are neither than activated relations. Drawing on our interpretive framework, integration is considered as a dynamic process, being not limited to a mere structural connection, but implying a synergistic action based on of win-win logic.

The structural connection expresses a condition of relational consonance, according to which subjects establish between themselves an infrastructural or communicative bridge that fosters the emergence of a mutual dialogue. However, to be sure about the existence of relations, it is necessary to prove if and how the related infrastructure is used. According to a structural perspective, the emerging of non-occasional interactions (e.g. not based on exchanges justified just by a simultaneous and short-term convenience) is considered desirable. Indeed, in this case, they tend to assume the form of a transaction and not of a relation. In fact, within relations, long-term value exchanges occur when resources are added to the process, nurturing a value offer (resource exiting in the system), which can be considered as an incoming resource. Therefore, relations are exchange interactions able to integrate resources. They also differ from competitive interactions, which are characterized by the lacking of exchange activities among actors, because each of them tries to influence the competitor in order to damage him/her or gain an advantage. Finally, relations cannot express binding links because of a dependence condition; thus, dependence forms tend to sooner or later collapse. Consequently, relational interactions are paths of synergistic resonance, characterized by a forward-looking perspective. Moreover, they are projected and stable over time, mainly because they are non-occasional (Vicari 1991) and expression of a high collaborative "commitment". Ultimately, none of those relations are based on a mere economic convenience and none of them generate (degenerate) specific form of dependence. Thus, they are always collaborative and dynamically resonant interactions characterized by a long-term perspective.

11.4 Diffusion and Definition

Maturana et al. (1985) stated "Everything is said by an observer". The above-mentioned expression assumes particular relevance in system thinking studies, implying a fundamental assumption according to which system is neither an organic and conceptual construct nor an objective and realistic representation of reality, being, on the contrary, a "way of observing".

According to our view, firm is not the observer (Vicari 1991), but it is its "central" component, considered as singular or plural sub-system organizational units.

From its (active and creative) observation perspective, the *Ego*, lacking of its own resources (except for knowledge and competences), considers its interlocutors (*Altera*) as (tangible or intangible) resource holders, useful or necessary to nourish the complex process of value co-creation that gives meaning to the emerging system. Hence, the specific context represents that portion of environment that the observer perceives as directly belonging to his/her firm (Golinelli 2010), being animated by a number of interlocutors, who participate in developing a complex relations' network aimed at exchanging resources (Barile and Polese 2010). Drawing on previous consideration, ignoring the rapports of dependence because of their inherent pathological nature that depends on the degree of mutual involvement, the exchange rapports can be considered as both transactions, when based on the research of short-term economic convenience, and reports, when based on a collaborative logic characterized by a strategic and potentially system projection. Some *Altera* are perceived as the most relevant and hopefully closer (to engage and emotionally retain); others are recognized as far as they are characterized by a predominantly utilitarian interest.

The separation among them cannot be imagined as a clear line of distinction, but rather as an area or a perimeter band characterized by high mutability.

The relational logic allows at the same time the gradual expanding, internalizing and integrating of the specific context, transforming stakeholders into co-makers and engaging them into a value co-creation process (Pellicano 1994; Normann and Ramírez 1998). The above-mentioned considerations make necessary to rethink firm's intangible borders, which are identified as a social system. The definition of system results from the formation of a boundary between the inside and the outside, which arises from the needs of the observer, who, when constitutes its systemic identity, has to necessarily distinguish it from the surrounding. Each identity presumes the formation of a difference that Luhmann (1984) considered "not ontological", but semantic because dependent on the operations (relations for resources/services integration) that foster the system in maintaining and reproducing over the time its reticular structure with respect to outside (autopoiesis).

The autonomy of the system from the environment (self-reference, self-organization and autopoiesis) can be explained assuming the vision according to which the system contains all the elements (resources and relationships) necessary for its survival (closed co-creation circuit).

The social systems organizational and functional autonomy is twofold:

- The aptitude of the system in conditioning the environment (no longer considered crucial);
- The aptitude of the structure (relational network) with regard to the decision maker influence (conscious guide), which is quite similar to what happens in biologic systems (particularly in human beings) when a relationship between mind and body occurs.

Figure 11.3 depicts the relationship among the firm and its environment. In particular, the general environment contains the specific context, in which are contained both the relevant *Altera* and the *Altera* characterized by utilitarian interests. The most relevant *Altera* are highly integrated in firm; thus, we represent them close to the observer, posed at the centre of the whole system.

According to the relational logic, a company draws on resources within its specific context (subjective), while the general environment (objective) plans it

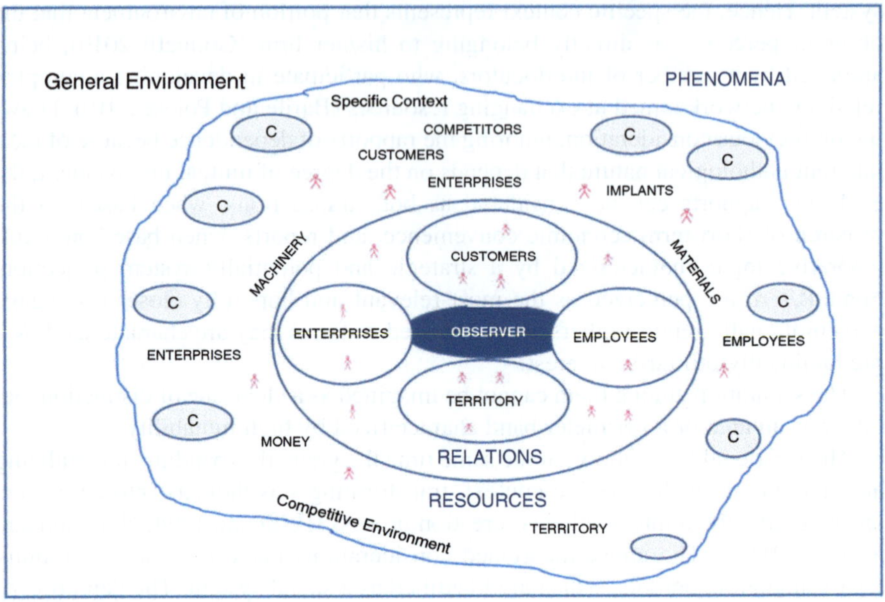

Fig. 11.3 The firm and the environment

on the Observer Subject (*Ego*) perception of threats or opportunities. Even in this case, it is important taking into account the real nature of relations which turn around company and can be potentially considered as a resources holder. However, on one hand, there are those who act according to an occasional need and to exclusively achieve a short-term economic convenience, without any particular interest or involvement in firm. On the other hand, there are those that aspire to cooperate showing their relational attitude. However, as argued below, the simple willingness does not identify the nature of relational network as systemic.

In this direction, the system integration, even if based on the sense of belonging, takes also into account the principle of accountability. Nevertheless, it is not a unidirectional responsibility of company, for example with regard to social stakeholders, but a reciprocal responsibility. Therefore, if there is a responsibility of the *Ego* towards its interlocutors, there should also be a responsibility of the interlocutors towards the company and its governance. If an interlocutor (*Altera*) shows a cooperative attitude, he/she is open to support the company towards a more effective, efficient and sustainable development. In sum, it allows company to be more viable and to nourish the conditions needed for reiterating over the time synergistic value creation processes. Hence, the subjects related to the observer (e.g. customers and suppliers) can be considered "part" of the enterprise system (Barnard 1939) only if they develop an intense and responsible sense of belonging (which respects the principle of commonality toward a purpose); thus, they link the satisfaction of current and future expectations (influence) to the viability of the system itself. However, they can be considered as "components" of the enterprise system not as people, entities or financial resources holders (skills and expertise become "services" in trade), but as relational *Altera* of the Observer Subject (*Ego*). In other words, their belonging to the system depends on the loyalty and the disposition in mutually feeding the relation.

The clusters indicated in the figure do not refer to all actors, but just to loyal ones. In terms of companies cluster, some insights arouse from the empirical evidence found out in local production systems, characterized by extensive and pervasive cooperative dynamics, which, in the Italian model, led to the conceptualization of the fourth capitalism (Bonomi and Rullani 2005).

The inclusion of the system theories in business and management studies is currently central in systemic sociology (based on the approach of Luhmann), making it possible to overcome the traditional concept of "organized structure."

The structure from which the system emerges is relational and reticular, whereby it is possible to consider company as a network of relations in action, emerging from the structuring of communication flows. Its immaterial border should be read as a mutual and coincident (between *Ego* and *Altera*) membership perception (Pellicano 2002), e.g. considering the co-maker (*Altera*) role in value co-creation processes driven by the Observer Subject (*Ego*).

11.5 Sense and Cohesion

Social systems are interrelated with biological systems, but differ from them because the former are established and organized according to the "sense" (Luhmann 1984). Even if individual players are biological organisms, social systems are mainly presented as non-organic entities, held together by symbolic processes, capable of providing decision-making information and mutual orientation criteria.

Drawing on the above-mentioned considerations, we define the sense as a "symbolic resource" that makes the mutual understanding and communication (relational) possible and plausible.

When an order or social cohesion does not realize, the actors perceive symbolic resources as not stable forms of meeting or communication (relations) capable of making comparable mutual expectations. When forms of meeting or communication do not arise, each actor (*Ego* and *Altera*) maintains his/her own expectations, depending on the contingency of the moment or characterized by a psychological unintentional nature. This is known as "double contingency mechanism" (as mutual between *Ego* and *Altera*) described by Parsons (1951), which has been overcome by the "symbolic media of interchange"—e.g. a common "sense"—essential to ennoble a social system.

The media communication aims at reducing the double contingency. The symbolic media, among which Parsons places money, influence, power and affection, play the role of arranger, establishing and codifying those expectations, which can become reciprocal and, consequently, relational.

According to the relational logic and in light of Service Science (SS) (Spohrer and Maglio 2008), the media or fundamental communicative code is identified in the "service", whereby value co-creation represents the "common sense" that ensures the social cohesion within the system. In other words, the viability of service systems depends on the ability of its governance to create and develop mechanisms for value co-creation, based on a continuous mediation of stakeholder's expectations (Spohrer et al. 2008) in a service logic, which represents the application of skills of a part for the benefit of another (Vargo and Lusch 2004).

Hence, in a broad sense, the function of codes such as money, power, scientific truth, service, love and so on, represents the technique that makes (relational) the communication accessible and available, fostering the formation of a non-ephemeral social order in economy, politics, academic life, personal and business relationships.

A system is characterized by the ability in maintaining a social order, possible thanks to a "glue" that puts and keeps together the different elements. Many authors focused on this "glue" highlighting different positions. According to Durkheim (1893), the social order of the systems is guaranteed by "solidarity", which is a rule set antecedent to contract rules; on the contrary, Hobbs (1954) and the English utilitarianism school stated that the this "glue" can only be represented by "personal interest".

We believe that a "sense" system arises when its members are able to give a shared meaning of their interactions; thus, the relational view is specifically based on for the strive to achieve a common sense. According to system theory, in a company the glue should be assured by the basic elements, such as values, shared rules, utilities and interest.

Actors can have relations because they deeply feel the ethical value of sympathetic reciprocity, follow rules, or (above all), finally, believe they can better meet their specific interests by taking part to an interpersonal relations system. The common—and systemic—purpose cannot be a mere sum of the interests of each component of the system, but what gives a meaning to a common interaction is a totally shared aim, the system viability and survival.

According to the relational logic, the rising of a common sense among participants (*Ego* and *Altera*) can be read in the light of autopoietic processes of value co-creation and resource regeneration. Whereby, subjects are joined together because they consider themselves as co-makers and, consequently, an active part of the interaction processes in which they participate, at the same time, both as suppliers (that is bearers of own resources) and as users of the new resources rising from systemic interactions. They are conscious that, if they predispose themselves to actively and responsibly participate in co-creation processes, they can also contribute to generate a value which they benefit in order to obtain a greater utility. The joint action directed by a shared sense (co-creation) develops a systemic membership that led *Altera* to feel more and more connected to the system as integrated parts. Membership, which ultimately induces loyalty, serves to better engage *Altera*, gradually increasing the quality of relational balance (Barnard 1938), in terms of contributions-rewards. It is quite simplistic (i.e. more realistic) to generally believe that the development of membership is mainly based on individual utilitarianism (tangible and intangible benefits) and not on the responsibility towards society (as enterprises are social commons). However, these reflections need to be read according to the greater or smaller geographical spread of shared ethical values of social responsibility.

Therefore, the "widespread engagement" (Freeman et al. 2007) among the interlocutors is possible only if a high adhesion and cohesion is developed among them, implying both emotional aspects (solidal responsibility) and utilitarian conveniences (benefits). Therefore, the participation is due to both "love" and "calculation" (Pellicano 1994).

11.6 Decision Formation and Formulation

Decision Making process is usually defined as a complex and dynamic process aimed at elaborating, focalizing and formalizing a potential strategic orientation (Casali et al. 2016). However, according to our view, this definition needs to be enhanced by the influence of relational logic. In particular, in Decision Making is possible to define two main moments, such as:

- "Formation" of strategic choices, the synthesis of which is represented by the focus of visions emerging from an inter-relational and interdependent dialectic (Game Theory): these visions should be implicitly shared among all members of the enterprise system, considered relevant by the *Ego*. Therefore, the strategic elaboration is ultimately the result of a Multiple Decision Maker (MDM).

  In forming decisions, the interest of recipients, for whom decisions are taken, is important. In this sense, firm can be considered a "nexus of interests" since interlocutors (*Altera*) of "Decision Formulator" (*Ego*) completely manifest their nature as stakeholders and their influence over system, if conceived according to a viable system view (see Fig. 11.4). However, a conscious diffusion of decision-making power should be supported by an equally widespread assumption of responsibility, related to the destiny of enterprise system.
- "Formulation" of the most important strategic choices of the *Ego*, to which is reserved the role of main actor of the whole decision-making process. This moment represents the end of an articulated dynamic characterized by an *Ego* responsibly, engaged in researching a better-weighted consensus or systemic resonance. This requires the constant coordination of negotiations among parties, bearer of their legitimate interests and of the subsequent expectations to search for a synthesis which, in accordance with the specific relational balances (contributions vs. gratification), can determine the maximum possible degree of systemic resonance.

Therefore, we can state that the systemic resonance represents the synthesis of two fundamental drivers: (1) the relational consensus expressed by the *Altera*, and

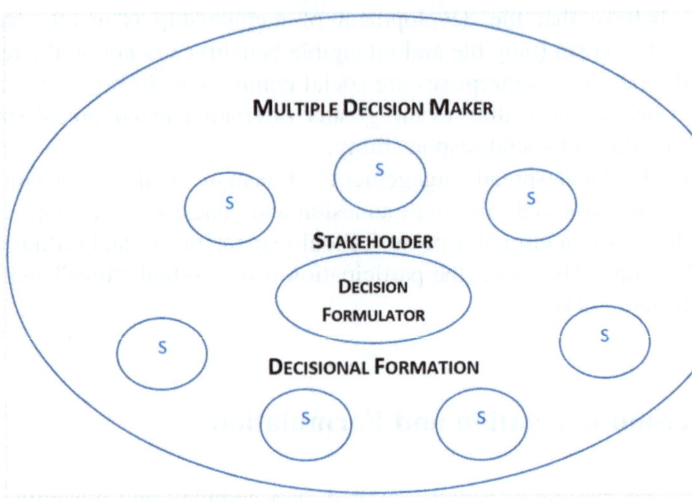

Fig. 11.4 Relational decision making process

(2) the perceived relational relevance attributed by the *Ego* to each cluster of stakeholders. The weighting of the first driver respect to the second one leads to reflect about how the search for consensus of the *Ego* is not absolute, but relative, because related to the search for the best balance among the competing interests that express the level of relevance perceived by *Alter*.

In summary, the *Ego* constantly looks for the consensus of *Altera*.

The final step is the focalization of the vision, consisting in its formal deliberation from which the resulting responsibilities of the formal governmental entity come out.

The corporate governance of a relational firm definitively corresponds to the described coordination and decision-making orientation activity.

11.7 Co-creation and Regeneration

Service Science (SS) points that, according to a relational logic (and in a network), resources are not something to "exchange" but rather something to commonly "use". In this regard, it is important to remember that SS is built upon the Service Dominant Logic (SDL) which provides a new way to look at the world in terms of entities (resource integrators), which normatively ("lawful") interact to co-create value (Vargo and Lusch 2004). SDL's eleven foundational propositions begin with the premise that service is the fundamental basis of all the exchanges. Consequently, SS defines the service as a value co-creation phenomenon that occurs when service system entities interact according to those value propositions that drive the application of competence for mutual benefit. Consistently, the

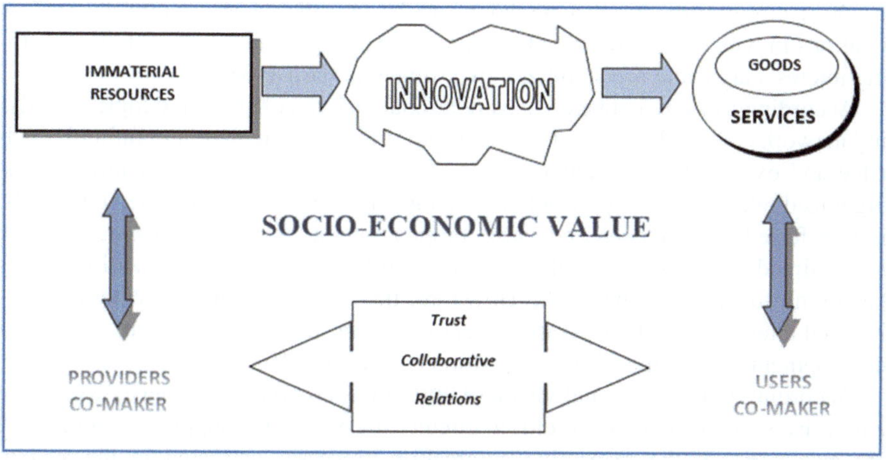

Fig. 11.5 Systemic value co-creation. Source: Pellicano (2002)

autopoietic nature of firm (social organism) leads to believe that internal relational processes (self-referential from a purely operational point of view) determine the "resource" regeneration. Briefly, in activated relations network, resources are integrated to generate new resources (Vicari 1991).

In light of this, enterprise system can be read as a complex bundle of processes which, starting from the subjective resources integration, co-create, as a result of inter-subjective actions, value in its plural (social, economic, equity, etc.) dimension, in the form of service proposition conceived as knowledge, skills, abilities and availability exchange.

Hence, the system stays viable thanks to an appropriate level of satisfaction (of their expectations) and the consensus of the various related (more or less relevant) subjects; thus, such a consent allows continuous resource feeding and regeneration, conceived as viability nourishment (see Fig. 11.5).

The interpersonal dynamics of creation and regeneration, now supported by the ICTs typical of the "social economy", represent the "sense" of being together of relational enterprise actors and constitute a closed circuit, since everything is inside the system (Pellicano 2002).

The emerged considerations, according to which value co-creation feeds collaborative and trust-based relations, lead to reflect about the progressive dematerialization of the processes that generate value.

The phenomenon originates from the economy of immateriality, which incorporates both the factors of production and the value proposition of company that tends to absorb not only a growing amount of scientific and technological knowledge (Rullani 1989), but also a relational dimension even more important in terms of economic value (Ciasullo 2010).

On a closer inspection, a network can be considered as a control mode of relations in which cooperative interactions, driven by common goals and converging interests arise and are stabilized over time (Barnard 1938).

According to a competence-based relational vision, collaborative logic underlying networks precisely responds to the need of aggregating and integrating knowledge and expertise. Thus, enterprise is conceivable as a network of mostly self-organized relations through which knowledge, dynamically transformed into skills, is transferred in processes and value propositions as part of a learning network logic, aimed at enhancing intelligence in network involving all organizational units belonging to the value network. Therefore, this vision, grounded on know-how-induced interaction and cooperation processes, is far from an interpretation of skills development based on endogenous efforts (Ciasullo 2010).

According to the relational vision of enterprise, co-creation exhausts within the enterprise system (in a sort of closed circuit) since resource suppliers (co-makers) are also users of the respective co-creation processes activated in the system (Normann and Ramírez 1998).

The development of this perspective, summarized by the relational logic, was originally (Pellicano 2002) focused on the following aspects:

- The dematerialisation of business economy, for which service logic becomes dominant in any business, market and society;
- The importance of intangible resources, linked to skills and competences, built together with stakeholders and not gained from the market, in accordance with the well-known passage from the property to the resource availability;
- The emphasis on contextual relationships aimed at developing the shared capital and enterprise knowledge capital;
- The participatory and responsible involvement of stakeholders in the processes of value co-creation even in their direct interest;
- The qualification of the generated value as a complex of specific value propositions in the plural, social and sustainable economy;
- The crucial role of trust in relation reproduction (Vicari 1991).

The described ideal-typical model of relational enterprise, essentially in line with the assumptions underlying the SDL (Vargo and Lusch 2004), counteracted with the ideal type of 'transactional enterprise', which is based on the exchange of material resources (G-Logic), realized through transactions and negotiations driven by a short-term economic convenience, grounded on the power (Pellicano 2002).

Therefore, the founding element of relational logic is the transition from the specific context to the relations (see Fig. 11.3) in the inter-subjective relationships among enterprises. The relational level highlights the need to go beyond "one-shot" exchanges (Reficco and Vernis 2010), typical of a conventional market logic. The B2B (Håkansson 1982; Morgan and Hunt 1994), marketing report (e.g. Sheth and Parvatiyar 2000), service (Grönroos 1991; Gummesson 2002; Gummesson and Polese 2009), interaction and network approaches (for example, Håkansson and Snehota 1995) and SD Logic (Vargo and Lusch 2004; Lusch and Vargo 2006) converge toward the need to overcome the traditional transactional logic, to adopt what we define the relational logic, which extends its application to all business relations existing between the Governance Entity and its interlocutors. This implies an extension of marketing, whereby: "Marketing has to change because society is changing" (Fabris 2008). In such a direction, firms should develop a continuous dialectic confrontation able to exceed the reference market to spread toward the broader society.

It is suggested that an important setting for the relational logic can be envisaged in the logic of value co-creation related to the concept of "many to many" defined by Gummesson (2004) that highlights some significant features. Hence, according to the relationship marketing and SD Logic, the many to many describes a model (meta-system) based on the meeting/collaborative and co-creative comparison between two systems: on the one hand, the supply system (in other words, the system of several service providers); on the other, the system of demand (i.e. the system of several service users). As mentioned before, the innovative element of this perspective consists in the search for a collaboration between the two systems aimed at increasing synergy satisfaction of mutual expectations. However, what the relational logic wants to highlight is that in the relational enterprise logic the active relation network leads to the emergence of a unique system among all related subjects.

Indeed, the "many to many" logic, reinterpreted according to a relational view, considers both many resource suppliers and many value propositions users as co-makers of a single systemic co-creation dynamics, although divided into several processes. More precisely, the co-makers of a closed-circuit co-creation logic, play, at the same time, the role of suppliers, resource holders, users and recipients of the value involved in the offers directed to them, which they have contributed to create and which they conjointly assume the responsibilities for.

The relational dynamics based on "many to many" logic echoes the concept of "ecosystem services", introduced by the SD logic and understood as "self-sufficient and self-regulating systems including actors that integrate resources linked by shared institutional logics and mutual value creation through service exchange" (Vargo and Lusch 2011; Lusch et al. 2016).

Outside enterprise system, there are other systems with which there is no exchange relation and co-creation activity. In particular, it is possible to identify:

- A general environment systems which generate influences (i.e. threats and opportunities);
- Other enterprise systems which, when there is a high degree of commonality of relevant resources, engage in a negative conditioning of competitors' relational skills.

According to the above-mentioned considerations, it is possible to state that SS and the relational logic have different purposes and exploratory fields:

- Service Science adopts an "objective" observational perspective, describing socio-economic macro phenomena such as the ecosystems in which actors (according to the many to many logic service systems such as customers,

suppliers, bidders, employees and so on) live together, collaboratively and co-creatively interact.
- The relational logic adopts a "subjective" and perceptive observational perspective, that is the vision of a subject defined *Ego* (entrepreneurial) which, by observing the surrounding environment, extracts and defines its specific scope (or resource field) and its relational system (with *Altera*).

The enterprise system, resulting from the central subject (*Ego*) perception, obviously appears immersed in a general ecosystem.

11.8 Resource and Competitiveness

Enterprises, being basically viable social organisms should feed constantly in order to survive, as like as any other living biological organisms.

The nourishment arises from material and immaterial resources that *Ego*'s interlocutors (*Altera*) hold. Companies are involved in a competitive struggle to gain these resources, which are scarce and strategically relevant. To this end, the *Ego*, in this case considered as decision maker, promotes and develops an "organizational action" (Thompson 1967; Weick 1968) consisting in forming and cultivating the "relational network", which precisely represents the organizational pattern of enterprise system.

A better quality of relations fosters the reduction of the "environmental" dependence (Salancik and Pfeffer 1978) or the negotiating power of the firm actors, who, sharing a common sense making (Weick 1995), develop a high engagement and sense of belonging to the system.

To re-read competitive dynamics according to a relational view, it is possible to refer to two concepts:

(a) The relevance of interlocutors;
(b) The commonality of interlocutors.

Two attributes characterize the relevance of interlocutors: (1) the criticality, e.g. the importance of the resources that the interlocutor held and allocate; and (2) the influence, expression of the interlocutor's ability in acquiring and re-appropriating resources, which defines restrictions dictating behavioural rules directed to the firm. In other words, the influence represents the interlocutor ability in appropriating adequate value margins, e.g. a highly positive balance in terms of contribution to value creation.

In sum, the criticality of relationship is due to all possible solutions to a problem and to the set of all the components needed for solve the problem itself. In particular, there are two types of criticality:

- Relation criticality;
- Resource criticality.

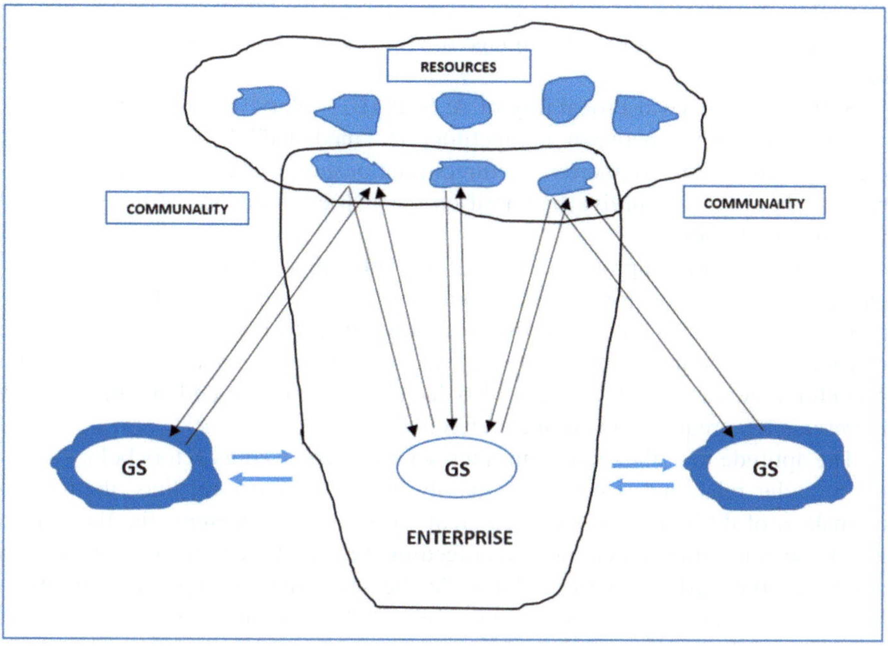

Fig. 11.6 Competitive dynamics

It is possible to state that the criticality represents the eliminability degree of an interlocutor, minimal when both the criticalities are maximum (see Fig. 11.6).

The commonality of interlocutors gives a sense to their perception, making them significant. On a closer inspection, this characteristic is considered innovative, because its application to the definition of potential competitors, being the most of them perceived as relevant interlocutors, having the higher degree of commonality. Therefore, the degree of commonality is directly proportional to the number of common interlocutors.

The competitiveness of relational firm would be rooted on interlocutors and on their ability to be more satisfying, than other competitors. Therefore, the competitive action needs for competitors' resources, affecting their ability in surviving and thriving.

The ability of competing companies to have "relatively" satisfied interlocutors is directly related to the existence of common relevant interlocutors; thus, this ability can be expressed by the following equation (Della Piana 2010):

$$\frac{S\,(A)}{S\,(B)} > 1 \qquad V.C.\,(A) > V.C.\,(B) \quad PS\,(A) > PS\,(B)$$

A and B = Two companies that have common relevant interlocutors;
S (A) = Satisfaction of stakeholders common to A and B respect to company A;

S (B) = Satisfaction of stakeholders common to A and B respect to company B;

C. A. (A) = Competitive Advantage of A; C. V. (B) = Competitive Advantage of B;

S. P. (A) = Survival Probability of A; S. P. (B) = Survival Probability of B.

The relationship between competitors is "mediated" by the interlocutors' (*Altera*) satisfaction; so, there is no direct relationship of exchange between competitors, because the survival is based on the indirect influence of the reciprocal viability conditions.

According to this approach to competitors' identification, the degree of dangerousness of the competitors would be precisely explained by the degree of commonality of relevant interlocutors. It follows that two or more firms can be considered competitors if they share, at least, one interlocutor perceived as relevant. Therefore, the survival of system is also due to the aptitude to collaboration, which represents a prerequisite to compete effectively.

The aptitude to collaborate with interlocutors, perceived as actors belonging to the reticular relational structure, is mainly due to their nature; thus, their being co-makers of differentiated value co-creation processes represents the basis upon which the ability to compete (aimed at feeding the vitality with the resources) with other entrepreneurial systems is built. At the same time, competitive successes reinforce the sense of belonging and the collaborative intention of the actors belonging to the entrepreneurial system.

Metaphorically, profitable partnerships and competitive capacities represent the two pillars, which dynamically allow company to develop paths of viability (Pellicano 2002). Thus, the competitiveness of a company depends on the resources available in a given environment and on the Governance Subject's ability to look for those resources needed for nourish the firm. These latter resources are limited and, therefore, strongly disputed. In other words, the resources are not sufficient for all that need or desire to achieve them. For this reason, the firm *Ego* looks for *Altera*, e.g. interlocutors with whom to establish relations.

Following this perspective, firm ultimately made up in decision maker and its interlocutors, the resource holders involved in value co-creation processes.

Such a competitive dynamic is the "communality"; in other words, competitors are business companies that share the same resource power. The more the community is strong, the greater is the competitive pressure exerted by the competitor perceived by the enterprise *Ego* as "direct" and therefore more dangerous. The relationship existing among competing firms is not based on exchanges, but on a negative mutual conditioning.

If resources are not enough to feed all systems, the interactions can be either competitive or cooperative. Hence, the competitive game becomes a struggle for survival.

Ultimately, relational firm, thanks to its disposition towards the development based on the integration with its own context, gains a higher competitive advantage and higher viability, as well as the ability in dealing with the daily struggle for material resources and intangible assets, which are limited and fiercely disputed (Pellicano 2002).

11.9 Leadership and Viability

Enterprise system, considered as an integrated relation set, requires some sort of "coordination of behaviours" (Vicari 1991) which can be partially accomplished by *Ego's* enterprise. Moreover, just in simple business realities, the (substantial) decision maker coincides with the (formal) *Ego*, e.g. with the top management (single, chief and board) which is a direct expression of the ownership. In complex socio-economic systems, the role of a decision maker (*Ego*) is quite complicated, because of the potential presence of numerous organisms that can still play a role of influential advisers, despite they not have a formal role in the organisational governance. Therefore, the relational logic requires a conceptual review of firm leadership view, whereby a relevant action of the decision maker is the communication. The ability in communicating is expressed in the aptitude to cultivate relations with interlocutors-co-makers, facilitating largely self-regulated value co-creation processes and fostering the autopoietic resource regeneration feeding the viability of the system and its consequent survival (Pellicano 1994; Pellicano and Perano 2007).

The decision maker is responsible for the harmony, ensuring firm survival, but it cannot be considered the only responsible for this success. The role of decision maker "should be considered a basic strategic guidance, a language, a culture, without planning, but communicating and involving"; thus, the decision maker has to learn how to facilitate interpersonal relations, fostering communication without directly managing it (Pellicano 1994).

The central element of an enterprise system is the decision maker, because of its leading role. However, today leading a company, understood as sense relational system, assumes different meanings. According to our view, the decision maker should play the role of relational enabler or facilitator. From its observational perspective, the decision maker continuously oversees and controls the evolutionary dynamics, knowing that more interactions smoothly develop, more the company keeps in harmony, e.g. in optimum condition to regenerate the conditions of its viability. Therefore, the relational logic, in harmony with the theory of viable systems (Beer 1989), enhances the role of the *Ego* in useful and well-defined reading key.

The awareness about the decision maker harmonizing function—in the fundamental relation among firm structural aspects and the environment in which it is immersed (Brunetti 1997)—is in line with the evolution of business strategies studies (Mintzberg 1985).

The proposed systematization contributes to identify the survival as a cause and, at the same time, as an effect of circular and cyclic processes of resource incorporation and skill and knowledge enhancement, being the incorporation the logical starting point for renewing the resources released by the counterparts.

The cycle of entrepreneurial viability (Pellicano 2004) starts from the provision of basic capabilities created into the enterprise network. Through strategic and organizational learning processes, the interaction ability arises from many actor-provider's ability in generating knowledge (Senge 2006) that, if properly exploited, brings out the distinctive competencies on which building up a non-ephemeral competitive advantage (Prahalad and Hamel 1990).

First of all, this advantage is the result of the consensus expressed by actors-users, who, in the described dynamic, are not mere recipients of company's value proposition, but co-makers of value, co-responsible for the satisfaction of their own needs and expectations. Furthermore, it should be considered that users' consent (resource holders) always arises from a comparative assessment that they express comparing the enterprise value proposition with those of direct competitors. Getting a non-occasional consent is a prerequisite that leads the decision maker's interlocutors to release their resources to the firm. The stabilization of relations, thanks to the loyalty of sub-system actors to the system (depending on the viable system perspective), allows the incorporation of new resources in the enterprise structure, which become available for the decision maker's guidance action. In this way, the cycle of viability is virtuously concluded, allowing the regeneration of the structural set of basic capacities. The relational logic also involves an extension of the "satisfaction" view, which tends towards an overall relational dimension. Consent must be differential, in order to lead the satisfied interlocutors to release resources; in other words, the value proposition has to be perceived not only adequate to the expectations, but the best if compared to those of competitors. Only capturing this perception of relative greater satisfaction, that is the differential consensus of the interlocutor, it will be possible to get resources. The released resource, as long as not trivial, should be incorporated; thus, according to a relational perspective, this means that interlocutor's loyalty arises from the stabilization of the underlying relation. The methods to stabilize a relation go beyond the use of monetary hard levers, creating emotional and sentimental ties consistent with service experience (see Fig. 11.7).

The establishment and strengthening of such ties act as a deterrent to potential unfaithfulness of co-makers and should build a barrier to competitors' disruptive actions, which have a high influence on the dynamics of value co-creation. This happens because they plan to get the differential consent of these relevant subjects considered relevant in value co-creation dynamics for the quality of the resources they held resources.

The activated and stabilized relations represent the barriers to competition, e.g. actions pointing to avoid that systems of competitors could obtain the advantage related to the acquirance of the same resources. Concluding, the competitive advantage is buildable and defendable only if the cycle is closed and constantly regenerated.

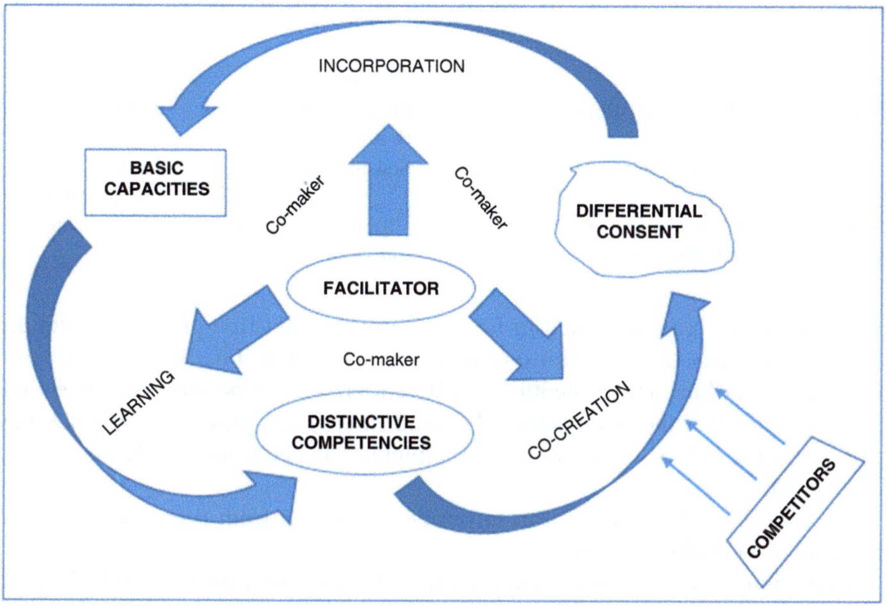

Fig. 11.7 Viable resource regeneration

11.10 Conclusions

This chapter highlighted the existence of seven possible approaches to the relational logic, which cover multiple facets of this topic. First, the concepts of *relationality and autopoiesis* refer to the relational nature of communication processes that constitutes and supports decision makers, being considered as an autopoietic social system. Second, *dissemination and definition* contribute to better understand the emergence of a system from the structure and as a result of the relational action of the *Ego* in the specific context, which is a "relative segment/divide" of general environment (ecosystem). Third, *sense and cohesion* consider a company as a system with a symbolic sense, represented by value co-creation processes and a sense of belonging arising from a complex mix of rational and emotional factors. Four, *decision formation and formulation* reflect problems' complexity, clarifying that decision-making seems to be even more complex, being the decision power shared and between *Ego* and *Altera*. Fifth, *co-creation and regeneration* delve with the value co-created within the relational network and representing the organizational pattern of system, which allows through the autopoietic regeneration of resources the flowing and fostering of its viability. Sixth, *Resources and competitiveness* describe the firm competitiveness as a system linked to the ability in acquiring resources thanks to the establishment of collaborative relationships. Seventh, *Leadership and viability* look at the decision-maker as the guarantor of the relational harmony, characterized by environmental dynamism. Therefore, its

Table 11.1 The roles that actors play in the three moments of entrepreneurial dynamic

Moments/ actors	Definition	Decision	Realization
Ego +	Observator/builder	Focalization/ (Formulation)	Facilitator (guide)
Altera =	Relational interlocutors	Stakeholders (supra)	Co-makers (subsystems)
Processes	Enterprising (systemic building—constitutive)	Visioning (decisional formation)	Acting (value creation)

constant control is fundamental in facilitating communication processes, with the ultimate goal of making the firm system able to stay viable. Finally, as results from the discussion of the above-mentioned different views, it is possible to connect the different roles that the main actors (*Ego* and *Altera*) can play with the three key processes of firm life arising from actors' interactions: constitutive, decisional and value creation.

As showed in Table 11.1, firm governance emerges from the joined dynamic made up of visioning and guidance.

Even though the relational logic is unified, it cannot be attributed exclusively to the decision-maker, being characterized by two meaningful components. First, the focus of strategic addresses that involves both the *Ego* and *Altera*; second, the exclusive role of the decision-maker as the responsible part involved in the relationship.

In conclusion, all the topics previously discussed contribute to define the foundations for the Enterprise Relational Vision (ERV) that emphasizes the role of firm's relationships in better managing the heterogeneity and complexity of the current business environment (Pellicano 2002).

The systematization of the ERV offers some practical and theoretical implications. The theoretical implications enable the enrichment of management literature, supporting new insights for the pre-existent relational view (Donati 2004; Bruni and Zamagni 2004; Pellicano 2002) and for Service Research theories (Vargo and Lusch 2004; Spohrer et al. 2007; Barile et al. 2010; Wieland et al. 2012).

In terms of practical implications, the definition of an original interpretative model let to clarify the dynamics within an integrated relational system. The interpretative model can be also seen as a suitable guideline for the governance, extremely useful for the decision makers.

Lastly, this chapter represents an initial contribution to the understanding of the broad and complex research area focused on relational logic, providing seven pillars. Further research is also needed to better examine those characteristics, analyze their possible connections and investigate other possible characteristics connected with relational logic.

References

Ashby, W. R. (1962). *Principles of the self-organizing system*. London: Pergamon.

Barile, S., & Polese, F. (2010). Linking the viable system and many-to-many network approaches to service-dominant logic and service science. *International Journal of Quality and Service Sciences, 2*(1), 23.

Barile, S., Spohrer, J., & Polese, F. (2010). Editorial column – System thinking for service research advances. *Service Science, 2*(1–2), i–iii.

Barnard, C. (1938). *The functions of the executive*. Cambridge: Harvard University Press.

Barnard, C. I. (1939). *Dilemmas of leadership in the democratic process*. Princeton, NJ: Princeton University Press.

Bateson, G. (1979). *Mind and nature. A necessary unity* [Trad. it. (1984). *Mente e natura. Un'unità necessaria*]. Milano: Adelphi.

Beer, S. (1989). *The viable system model*. London: Wiley.

Berry, L. L. (1983). Relationship marketing. In L. L. Berry, G. L. Shostack, & G. D. Upah (Eds.), *Emerging perspectives of services marketing* (pp. 25–28). Chicago, IL: American Marketing Association.

Bonomi, A., & Rullani, E. (2005). *Il capitalismo personale: vite al lavoro*. Torino: Einaudi.

Brunetti, F. (1997). Sull'architettura delle relazioni tra impresa e ambiente. *Problemi di gestione dell'impresa, 23*.

Bruni, L., & Zamagni, S. (2004). *Economia civile. Efficienza, equità, felicità pubblica*. Bologna: Il Mulino.

Capra, F., & Luisi, P. L. (2014). *The systems view of life: A unifying vision*. Cambridge: Cambridge University Press.

Casali, G. L., Franco, E., & Perano, M. (2016). *Strategic management. Growing into a foreign market*. Prahran: Tilde Publishing and Distribution.

Ciasullo, M. V. (2010). La competitività. In M. Pellicano & M. V. Ciasullo (Eds.), *La visione strategica dell'impresa*. Torino: Giappichelli.

Della Piana, B. (2010). L'analisi strategica. In M. Pellicano & M. V. Ciasullo (Eds.), *La visione strategica dell'impresa*. Torino: G. Giappichelli.

Donati, P. (2004). *Introduzione alla sociologia relazionale*. Milano: Franco Angeli.

Durkheim, È. (1893). *De la division du travail social*. Paris: Presses Universitaires de France.

Dyer, J. H., & Singh, H. (1998). The relational view: Cooperative strategy and sources of interorganizational competitive advantage. *Academy of Management Review, 4*, 660–679.

Fabris, G. (2008). *Societing: il marketing nella società postmoderna*. Milano: Egea.

Freeman, R. E., Harrison, J. S., & Wicks, A. C. (2007). *Managing for stakeholders: Business in the 21st century. Managing for stakeholders: Survival, reputation, and success*. New Haven: Yale University Press.

Golinelli, G. M. (2010). *Viable systems approach: Governing business dynamics*. Padova: Cedam.

Grönroos, C. (1983). *Strategic management and marketing in the service sector*. Helsingfors, Helsinky: Swedish School of Economics and Business Administration.

Grönroos, C. (1991). The marketing strategy continuum: A marketing concept for the 1990s. *Management Decision, 29*(1), 7–14.

Grönroos, C., & Gummesson, E. (1985). *The Nordic school of service marketing. Service Marketing–Nordic School Perspectives*. Stockholm University, 6–11.

Gummesson, E. (2002). *Total relationship marketing* (2nd ed.). Oxford: Butterworth-Heinemann.

Gummesson, E. (2004). From one-to-one to many-to-many marketing. In *Service Excellence in Management: Interdisciplinary Contributions, Proceedings from the QUIS 9 Symposium*, Karlstad University Karlstad, Sweden, 16–25.

Gummesson, E., & Polese, F. (2009). B2B is not an island! *Journal of Business & Industrial Marketing, 24*(5/6), 337–350.

Håkansson, H. (1982). *International marketing and purchasing of industrial goods an interaction approach*. Chichester: Wiley.

Håkansson, H., & Snehota, I. (1995). *Developing relationships in business networks*. London: International Thomson.
Hobbs, T. (1954). *Leviathan, or the matter, forme and power of a common wealth ecclesiastical and civil*. London: Andrew Crooke (1651).
Luhmann, N. (1984). *Soziale Systeme. Grundriss einer allgemeinen Theorie*. Frankfurt: Suhrkamp.
Lusch, R. F., & Vargo, S. L. (2006). Service-dominant logic: Reactions, reflections and refinements. *Marketing Theory, 6*(3), 281–288.
Lusch, R. F., Vargo, S. L., & Gustafsson, A. (2016). Fostering a trans-disciplinary perspectives of service ecosystems. *Journal of Business Research, 69*(8), 2957–2963.
Maglio, P. P., & Spohrer, J. (2008). Fundamentals of service science. *Journal of the Academy of Marketing Science, 36*(1), 18–20.
Maturana, H. R., Varela, F. J., & Frenk, S. (1985). *Autopoiesi e cognizione: La realizzazione del vivente*. Venezia: Marsilio Editori.
Mele, C., Pels, J., & Polese, F. (2010). A brief review of systems theories and their managerial applications. *Service Science, 2*(1-2), 126–135.
Mintzberg, H. (1985). Of strategies, deliberate and emergent. *Strategic Management Journal, 6*(3), 257–272.
Morgan, R. M., & Hunt, S. D. (1994). The commitment-trust theory of relationship marketing. *Journal of Marketing, 58*(3), 20–38.
Nahapaiet, J. E., & Goshal, S. (1998). Social capital, intellectual capital and the organizational advantage. *Academy of Management Review, 23*, 242–246.
Normann, R., & Ramírez, R. (1994). *Designing interactive strategy*. Chichester: Wyley, (trad. it.) Le strategie interattive d'impresa, Etas, Milano, 1995.
Normann, R., & Ramírez, R. (1998). *Designing interactive strategy: From value chain to value constellation*. Chichester: Wiley.
Pardi, F. (1998). *Teoria dei Sistemi, in Enciclopedia delle scienze sociali* (VIII, ed., pp. 19–26). Roma: Istituto Enciclopedia Italiana.
Parsons, T. (1951). *The social system*. New York: Free Press (trad. it.) Il sistema sociale, Edizioni Comunità, Milano, 1965.
Pellicano, M. (1992). *La comunicazione aziendale nelle imprese di servizi pubblici*. Padova: Cedam.
Pellicano, M. (1994). *Sistemi di Management*. Padova: Cedam.
Pellicano, M. (2002). *Il governo delle relazioni nei sistemi vitali socio-economici* (n. 13). Torino: Giappichelli.
Pellicano, M. (2004). *(a cura di) Il governo strategico dell'impresa*. Torino: Giappichelli.
Pellicano, M. (2005). Il governo delle relazioni d'impresa. *Sinergie, 21*.
Pellicano, M., & Perano, M. (2007). Esposito De Falco, S.: Analysis of the relational capital devaluation risks within the organization. In *Proceedings of 10th International Conference Society for Global Business & Economic Development (SGBED)*. Kyoto, 8–11 August 2007.
Pels, J., Barile, S., Saviano, M., Polese, F., & Carrubbo, L. (2014). The contribution of VSA and SDL perspectives to strategic thinking in emerging economies. *Managing Service Quality, 24* (6), 565–591.
Prahalad, C. K., & Hamel, G. (1990). The core competence of the corporation. *Harvard Business Review, 68*(3), 79–91.
Reficco, E., & Vernis, A. (2010). Market ecosystem and social inclusion. In P. Márquez, E. Reficco, & G. Berger (Eds.), *Socially inclusive business in Iberoamerica: Challenges and opportunities*. Cambridge: Harvard University Press.
Rullani, E. (1989). La teoria dell'impresa: soggetti, sistemi, evoluzione. In M. Rispoli (Ed.), *L'impresa industriale*. Bologna: Il Mulino.
Salancik, G. R., & Pfeffer, J. (1978). A social information processing approach to job attitudes and task design. *Administrative Science Quarterly, 23*(2), 224–253.
Senge, P. M. (2006). *The fifth discipline: The art and practice of the learning organization*. Broadway Business.
Sheth, J. N., & Parvatiyar, A. (2000). *Handbook of relationship marketing*. Thousand Oak, CA: Sage.

Spohrer, J., Maglio, P. P., Bailey, J., & Gruhl, D. (2007). Steps toward a science of service systems. *IEEE Computer, 40*(1), 71–77.

Spohrer, J., & Maglio, P. P. (2008). The emergence of service science: Toward systematic service innovations to accelerate co-creation of value. *Production and Operations Management, 17*(3), 238–246.

Spohrer, J., Anderson, J. L., Pass, N., & Ager, T. (2008). Service science and service dominant logic. *Otago Forum, 2*, 4–18.

Thompson, J. D. (1967). *Organizations in action: Social science bases of administrative theory*. New Brunswick: Transaction Publishers.

Vargo, S. L., & Lusch, R. F. (2004). Evolving to a new dominant logic for marketing. *Journal of Marketing, 68*(1), 1–17.

Vargo, S. L., & Lusch, R. F. (2011). It's all B2B... and beyond: Toward a systems perspective of the market. *Industrial Marketing Management, 40*(2), 181–187.

Vicari, S. (1991). *L'impresa vivente*. Milano: Etas Libri.

Vicari, S., Bertoli, G., & Busacca, B. (2000). Il valore delle relazioni di mercato. Nuove prospettive nelle analisi delle performance aziendali. *Finanza, Marketing e Produzione, 5*, 7–53.

Von Bertalanffy, L. (1968). *General system theory. Foundation development, application*. New York: Penguin University Books.

Weick, K. E. (1968). Systematic observational methods. *The Handbook of Social Psychology, 2*, 357–451.

Weick, K. E. (1995). *Sense-making in organizations*. Thousand Oaks: Sage.

Wieland, H., Polese, F., Vargo, S., & Lusch, R. (2012). Toward a service (eco) systems perspective on value creation. *International Journal of Service Science, Management, Engineering, and Technology, 3*(3), 12–25.

Wiener, N. (1948). *Cybernetics, or, Control and communication in the animal and the machine*. Paris: Hermann.

Chapter 12
From Mechanical to Cognitive View: The Changes of Decision Making in Business Environment

Mario Calabrese, Francesca Iandolo, Francesco Caputo, and Debora Sarno

Abstract Management can be considered the science of decisions. Different researchers and practitioners have analysed the contents and boundaries of this vibrant domain, proposing multiple contributions and several perspectives.

After a short overview of the principles of scientific management and their contribution to the definition of the modern idea of business management, this chapter briefly introduces the topics of subjectivity, complexity and uncertainty as relevant challenges to face in decision-making processes. The aim is to identify the key dimensions of the decision-making process over time.

Lastly, systems thinking, and specifically the Viable Systems Approach (*vSA*), is proposed as a relevant interpretative lens to overcome the limitations of a traditional reductionist view and better understand the challenges and dynamics of post-modern social and economic dynamics.

Keywords Systems approach • Decision making • Complexity

12.1 Introduction

Some might say that the information variety (or, simplifying, information heritage) of a viable human system (for example, a person or a society) is immune to sudden evolutions of society. However, if we analyse the story of humanity in depth, we can observe that great historical revolutions were subsidiaries of the adoption of new interpretative schemes of reality (the part of information variety related to the

M. Calabrese • F. Iandolo
Sapienza University of Rome, Rome, Italy

F. Caputo (✉)
Masaryk University, Brno, Czech Republic
e-mail: fcaputo@mail.muni.cz

D. Sarno
Department of Economics, University of Foggia, Foggia, Italy
e-mail: debora.sarno@unifg.it

S. Barile et al. (eds.), *Social Dynamics in a Systems Perspective*, New Economic Windows, DOI 10.1007/978-3-319-61967-5_12

ways in which we organise and represent perceived information), controlling the ascendant entropy in terms of increasing chaos (Barile et al. 2015a, b).

For example, the nomadic people of the past, who were dedicated to activities characterised by poor social structuring, such as gathering and hunting, had a different *modus operandi et pensandi* from those who lived in a formally structured society essentially based on agricultural activities. This phenomenon can be explained by the positive correlation between the improved quality of the organisation of society and a favourable change in the organisation of information flows. These variations in the exchange of information units affect the perception, understanding and organisation of reality in the viable human system.

Regarding the managerial sciences, the management of information flows and the collection, processing and interpretation of information units represent the theoretical essence worthy of attention from scholars interested in entrepreneurial dynamics. Notwithstanding, management (which can be referred to as the clarification and implementation of objectives through decisions) remained on the back burner until the emergence of the observation perspective of the classical economists, who considered the economy as an impersonal and objective science. Indeed, Kenneth Boulding Ewert (1961) affirmed that economic science is concerned with the behaviour of products rather than the behaviour of men. However, other thinkers, such as Saint-Simon (1760–1825), Say (1767–1832), Fourier (1772–1837), Shibusawa (1840–1931) and Towne (1844–1924) all emphasised the relevance of management.

The birth of the first managerial doctrine can be dated to the 1880s–1890s with the formulation of scientific management by Frederick Taylor. The word *scientific* underlines the extent of the American engineer's theories, in which the organisation of work is no longer based on rules derived from practice and intuition but instead on rules derived from science and technology and, therefore, from so-called scientific knowledge (Grönroos 1994). However, the Taylorist scheme considers the organisation as a formal structure that elaborates informative input through scientific methods, turning them into specific output: decisions (Merkle 1980).

Building upon these embryonic considerations, this chapter analyses the evolution of managerial approaches with the aim towards emphasising their functioning and contributions to facing entrepreneurial issues. Accordingly, the chapter is structured as follows: In Sect. 12.2, the principles of scientific organisations are analysed, and their contributions to changes of decision-making processes are presented; in Sect. 12.3, the role of subjectivity, complexity and uncertainty are analysed as key elements that affect the dynamics of decision making; in Sect. 12.4, the contribution of systems thinking, and specifically, of the Viable Systems Approach (*vSA*) in defining more effective approaches of decision making, is introduced. Finally, in Sect. 12.5, some conclusions and future directions for research are proposed.

12.2 Principles of Scientific Organisation

12.2.1 The Modern Industrial Era

Human history, including the history of management is a succession of events and facts in all the human fields, including the economy; therefore, the second Industrial Revolution can be counted among the great changes that have helped to radically and continuously transform habits. Many scholars talk about a great transformation (Angaroni 1977), as the revolution caused the abandonment of a slow and inefficient production system in which the population, output and income imperceptibly increased in favour of another in which growth was faster.

The Industrial Revolution also determined the birth of the modern organisation through the disruption of traditional productive processes. The steam engine became the emblem of this radical change (Debeir et al. 1987a, b).

Before the first Industrial Revolution, the concept of organisation did not exist: The relationship between master and worker was very simple and determined by centuries of constant practice (Mooney 1947), while the production process was not univocally defined and all workers retained a certain autonomy and, eventually, irrationality (Sombart 1969). Indeed, the importance of improving the labour organisation system was not recognised, and random or unexpected elements were managed time by time.

Although this was the era of *trusts*, each company was created and developed by the energy and skills of new people, usually men, who came from more or less humble positions and had experience as commercial or technical senior managers. Success was mainly due to complete knowledge of the entire industry obtained through personal contacts, and the training of subordinates was aimed at achieving maximum efficiency (Taylor 1914).

Corporate action was oriented not towards coordinating and connecting different activities, but towards decisiveness: 'the organisation is still considered a matter of men, according to the ancient belief that, when you have the right man, the methods can be safely left to his discretion' (Taylor 1914, p. 115).

This way of gaining knowledge and making the organisation advance had considerable limits, unless it benefited from the support of erudite and learned people (Winsor 1992). Mainly, it resulted in 'the lack of a clear relationship between a good industrial organisation and company profits' (Taylor 1914).

12.2.2 Scientific Management

Scientific organisation simplifies the activities carried out by man and, consequentially, enables them to be more productive. Taylor (1914) summed up his argument as follows:

- Replacement of workers' personal opinions with scientific knowledge;
- Selection and scientific training of the workforce so that each worker is examined and trained to perform tasks suitable for his or her skills instead of being left free to choose a job that improves his abilities;
- Intimate collaboration among managers and workers so that the latter conduct the work according to previously determined instruction obtained by times and methods studies, in contrast to a system which leaves the solution of any problem in the hands of the individual performer.

By applying these new principles instead of imposing an individual effort on every worker, the two parties share, in an almost equal manner, the tasks to be performed daily.

The principles of scientific management allow for the scheme of deriving knowledge after completing activities (trade logic) and encourage the use of *ex ante* knowledge as well as the assertion of rationality as the key logic (mechanistic logic) (Taylor 2004). However, over time, vibrant social and economic dynamics have underlined the uselessness of adopting ex-ante knowledge to face new (and unexperienced) problems and/or situations. Therefore, the will to abandon one paradigm and replace it with another emerges. The problem that arises at this point is that the decision of the scientific community to reject the old paradigm and adopt the new one is not justified by reasons of logic or an experimental nature; rather, it is the result of a conversion of a gestalt switch comparable to a political revolution, which changes the whole way of thinking about the identity of the business economy. In the Taylorist system, it is easy to notice the will to adopt the scientific method to allow, by overcoming the ingenuous pre-Fordism period, the progress of organisational thinking (Di Nauta et al. 2015; Formisano et al. 2015).

Aside from establishing the adoption of analytical reason, scientific management principles separate direction (management) from labour (workers), reflecting the mind–body dichotomy. In particular, the former has the responsibility to guide workers appropriately during the transition period—or, in other words, as the facts gradually replace the opinions. As Taylor also stated, the fundamental driver of the new paradigm is knowledge: In most cases (especially when the work being undertaken is complex by nature), the development of scientific knowledge is the most important of the fundamental elements of the new organisation. However, the new elements introduced by the new observation perspective do not automatically ensure coexistence with the pre-existing principles.

The image of the structure (machine) as the firm's representative model seems opposite to the previous one, which is grounded on the idea of patrimony of values. It becomes clear that management does much more than allocate values in accordance with the results of calculations (Saviano et al. 2016). Instead, like a mechanical engineer, management also has to design an overall structure (the organisation), ensuring that each part fulfils its function, and plan behaviours in a unified manner.

Summarily, scientific management is interested in the systematisation of the organisation of knowledge, the affirmation of scientific knowledge, and the monitoring of performance. In this type of organisation, the standardisation of methods is an essential prerequisite for both establishing the running time of each operation and ensuring that it is actually executed within the granted time limits (Maier 1970).

12.3 Beyond the Scientific Organisation

The need to include the human dimension in traditional managerial approaches and overcome the traditional and structural perspective, which were considered a limitation for organisations' plans and actions, has generated a new research stream interested in human relationships (Ramage and Shipp 2009). However, it should be noted that even though some researchers have considered the *theory of human relationships* as opposite to Taylorism, others (Anfossi 1971) have affirmed that it offers useful contributions to enrich Taylor's conceptual framework.

Indeed, by analysing the principles of rational economic theories, such as the maximisation of individual value, perfect competitiveness, and full knowledge of instruments to achieve planning aims, it is possible to derive the basis for a deductive science in which individuals' behaviours and choices are defined with reference to contextual elements without considering that, sometimes, they are not aligned with common and shared principles (Katona 1964).

Accordingly, the research stream on human relationships underlines the relevance of motivation and individuals' aims to the functioning and survival of organisations (Momigliano 1975). It highlights that the study of subjective variables (such as attitudes, motivations and frameworks of reference) is relevant to the understanding of production and organisational processes (Caputo et al. 2016a).

In the same direction, the research underscores the ways in which individuals' reactions to external pressures can change in accordance with their mental status or past experience and, for this reason, it is not possible to define a mechanistic approach to explain them (Sterman 2000). According to Emery (1981), to overcome the limitation of traditional approaches, it is necessary to develop new models and instruments able to support decision makers in understanding the relevant variables and the ways in which they affect individuals' behaviours.

From this point of view, an interesting advancement in knowledge was offered by Abraham Maslow, who, in his book, *Motivation and Personality* (1975), highlighted the impact of individual perceptions and mental models on the function of the organisation in which individuals act. As a consequence, this contribution changed the traditional hierarchy of Taylor's approach, demonstrating that individuals cannot be considered simple elements of an organisation (Senge and Sterman 1992).

The change in perspective proposed by Maslow's research established the conditions for increasing the attention paid to human resources; but, as a consequence, subsequent studies focused their attention only on individuals and their perceptions, not on the ways in which they interact and collectively influence organisations and their pathways (Maslow 1975).

12.3.1 The Role of Mental Processes

The executive manager Chester Barnard tried to mediate between the two predominant theses of the time: on the one hand, scientific management with its Olympic rationality; on the other, the school of human relations, with the primacy of human factors in the performance of organisations. Opposing Taylorism, in his work, 'The Functions of the Executive', Barnard asserted that the study of individual motives is essential for a full and proper understanding of organisations (Barnard 1968). These motives are defined as systems of activities or personal forces that are aware, resolved and finalised. Although not directly, Barnard mentioned the dynamics concerning knowledge referring to managerial issues. In particular, in a report presented to the faculty and students of the Faculty of Engineering at the Cyrus Fogg Brackett conference held on 10 March, 1936 at Princeton University, Barnard described the mental aspects of individuals in performing various and multiple tasks, including work and daily tasks.

Barnard distinguished mental processes as either 'logical' or 'not logical'. Obviously, in common experience, these two types of processes are not divided at all, but are rather an indistinct process—that is, knowledge. By 'logical processes', we refer to conscious thought expressed in words or other symbols—that is, through reasoning. By 'not logical processes', we mean those that cannot be expressed in words or as reasoning and simply express a judgement, with a concomitant decision or action (Stenning and Lambalgen 2005). This may be due to the fact that the processes are unconscious or, since they are so complex or rapid (often almost instantaneous), they cannot be analysed by the person in whose brain they occur (Marcel 1983). The not-logical thought processes range from no reasoned decision to *not putting your hand in the fire twice* to considering, in a flash, a mass of experiences or set of abstractions. We could not perform any work without these kinds of thought processes. Some of them are so inexplicable that we call them *intuitions*. Many go by the name *common sense*. Few are called *inspirations* or, sometimes, a *stroke of genius*. Most, however, are related to *meaning*, *common sense*, *judgement* or *a bright idea* (Lengnick-Hall and Lengnick-Hall 2002).

According to Barnard, logical processes are considered to be of a higher level than their not-logical counterparts since they allow us to rationalise reality, make action possible, and determine when actions are hidden or unconscious. This prevents us from understanding human behaviour and, consequently, from improving the functioning of organisations. Barnard did not want to minimise the processes of reasoning, but instead attempted to point out the importance of developing a new way of thinking, one which sees integrated and coordinated processes as logical *and* not logical (Gross and John 2003).

However, it is not easy to accept that managers follow feelings and give priority to first impressions, rejecting logical conclusions and, eventually, not undertaking meticulous analysis in favour of a comprehensive sense of a totality that involves an inconsistency of attitudes. To face this challenge it is need to developing artistic principles in the use of the mind and having achieved balance between speed and caution, among large patterns and the adequacy of details, and between strength and flexibility (Minsky 1988). Finally, Barnard aimed to investigate differences among the most common professions based on different attitudes in the required mental *composition*.

Two concepts developed by Katona (1964) and closely related to this stream are the following: authentic decisions and habitual behaviour. In particular, authentic decisions are not usual. They involve the perception of a new situation and the solution of a problem that is inherent to it; they urge people to react to a situation in a new way. The habitual behaviour, however, is much more common. In fact, we tend to do what we have already done previously in similar situations. It is inappropriate to use the word *decisions* for such circumstances: The psychological process in question does not involve authentic decisions. It is more appropriate, in this case, to talk about the behaviour of a *routine* or about the mechanical application of known rules. The difference between the two types of behaviour may benefit from further clarification by inserting the much-discussed problem of rigidity and flexibility into the construct (obviously, rigidity should be understood both in the global sense and in the molecular sense, and must be differentiated from impulsivity and fixation).

An example of rigid behaviour can be the cross-road that always leads to the house at a certain point. The way in which certain sellers set their pricing is another frequently used example of rigid behaviour. These are routine acts that are constantly repeated. In contrast, flexibility can be defined as a reorganisation of the field and an act within the framework of a new level of understanding. We can also use the terms *rational behaviour* and *irrational behaviour*.

These terms are also often used to indicate, on one hand, reasonable behaviour with a purpose and, on the other, emotional, causal, unreasonable behaviour. Understanding the differences in the various mental processes and the importance of not-logical thought processes would solve many problems affecting modern society. The fact that the ascendant entropy will always require more rigorous thinking skills is clear, but the result of simplifying and ordering the existing system will not be achieved without a corresponding re-evaluation of the not-logical mind. *Brains* with no *minds* seems to be a futile imbalance (Covey 1992). The methodological inconsistencies and the misunderstanding among large groups generated by increasing specialisation require the correction of the sentiment about the net balance and the interest of the whole and the spirit which, perceiving the concrete parties, also includes intangibles.

Following these reflections, it is possible to state that the uncertainty in terms of not imagined or not imaginable alternatives represents a consistent part of our everyday life. Building managerial approaches that do not take into account the role of uncertainty would mean acting without a compass. In the next subsection, a brief overview of the management of uncertainty is offered.

12.3.2 *Uncertainty in Making Decisions and the Role of the Organisation*

The introduction by Simon to the notion of bounded rationality summarized in the following Table 12.1 marked the end of certainty within manufacturing processes (Bonazzi 2002). From this moment on, uncertainty would be a constant in the development of organisational thinking. This statement is true with reference to two scholars: Thompson (1967) and Galbraith (1974). The distinctive aspects of both views are summarised in Table 12.1.

According to the contributions summarised in Table 12.1, the issue herein is part of the broader relationship between rational and natural models and, finally, of the

Table 12.1 Comparison of Thompson and Galbraith perspective about uncertainty

Object of comparison	Thompson point of view	Galbraith point of view
Definition of uncertainty	Further specifications of Simon's bounded rationality, in terms of analytical levels of certainty: – Internal level, referred to as the "technical core", which has the task of managing routinary operations and, therefore, works according to the principle of optimization (maximum rationality) – Institutional level, which separates the organization from the environment and in which adaptation and competitive behaviours are developed, aimed at survival – Managerial level, positioned between the previous ones, which acts as a filter with respect to the external uncertainties and as a regulator of the technical core, making it implementing the appropriate changes as a result of the changes that have occurred in the environment	Uncertainty is the gap between the information units within the organization and those needed to take a good decision (Galbraith 1973, 1977). Uncertainty is defined in negative terms as a gap to be filled
Cognition within the organization	The uncertainty of the environment, exerting different impacts on the individual components of the organization, determines inhomogeneity	Organization is a cognitive system that processes the information units. The cognitive effort is closely correlated to the degree of uncertainty
Ways to reduce uncertainty	Overcoming the rigidity of the industrial technology typical of mature Fordism, it is extensively introduced the role of technology, which includes mechanical equipment and devices as well as any specialized knowledge (Bonazzi, 2007). The non-perfection of technology and the dynamism of the context of reference emphasize the close link between knowledge and complexity	Two different strategies: – Ex ante reduction: simplifying the problems and adopting suboptimal solutions (Maraschini, 2002) – Ex post reduction: strengthening the adaptation processes to adapt to the requirements imposed by the context

distinction between mechanistic structures, organismic structures and cognitive structures. To analyse the topic of uncertainty, it is necessary to specify that several managerial contributions have emphasised that decision makers and practitioners can anticipate it (Ashley and Morrison 1997; Kandampully and Duddy 1999) and manage its dynamics by developing conditions of internal and external flexibility in order to ensure the survival of the organisation (Bracker 1980; Oliver and Holzinger 2008; Stalk et al. 1991). Despite the relevant advancements proposed by these and other contributions, we want to point out that at least a little bit of uncertainty always remains, and managers take this into account when they make decisions, not just relying on rationality. There is a need for new interpretation schemes for managing organisations that are able to both support managers in decision making and explain to researchers why decisions are made. Systems theories, and the viable systems approach in particular, significantly contribute towards overcoming the limits imposed by a low attention on the topic of uncertainty in decision-making processes.

12.4 The Contribution of Systems Thinking to Decision Making: A Focus on the Viable Systems Approach

12.4.1 The Shift from the Elements to the System

The mechanistic perspective, as stated, is based on a Cartesian view that provides an explanation of reality through knowledge of details. However, the statement of issues (such as the three-body problem, the uncertainty principle and the central role of the observer) affects the belief that an increase in the probability of exploiting the details can be transformed into knowledge and, therefore, into the governance of the dynamics of the external environment. Moreover, fragmentation creates a kind of general confusion of the mind, from which springs an endless series of problems, negatively interfering with our clarity of perception and preventing us from solving most of them (Bohm 2002).

The development of new forms of perception allows us to focus on the concept of interaction and, consequently, to shift the focus from the *elements* to the *systems* (Mella 2005). The emergence of systems due to the interaction between components is a representation of crucial importance in applications and theories. Thus, the study of such interaction has become *interdisciplinary* and, when it has been added to the interaction of different approaches and knowledge, has generated the *systemic approach* (Von Bertalanffy 1969). By adopting this approach, organisations should be studied like natural systems—that is, by abandoning the mechanistic–reductionist vision (Pascale 1999). The adoption of this approach, which involves the analysis of complex interactions between the organisation and the external environment, does not imply the isolation of the organisation itself on the conceptual level (Emery 1981).

12.4.2 Knowledge, Adaptation and Learning

Several later studies investigated issues related to the properties of social systems; in the following, the aim is to illustrate how knowledge has an impact on the conceptual frameworks developed by the social science interested in systems thinking. In this regard, we can refer to Ashby's precursory work, which emphasises the importance of requisite variety (requisite multiplicity): 'only variety can destroy variety' (Ashby 1956). In other words, to govern complexity, a system must have an internal variety (knowledge) consistent with the external variety (entropy, uncertainty).

In the struggle for survival, organisations necessarily evolve to address increasing internal chaos as well as the chaos derived from their relationships with other entities populating their context. Therefore, in order to understand the nature and evolution of organisations, it is necessary to focus on the theoretical aspects of complex systems. In this regard, requisite variety, understood not only as an adaptation to new situations, is a scheme that allows the analysis of the functioning of cybernetic models and their application to the business system. Specifically, Ashby (1960) underlined that the adaptation by learning method offers a system the opportunity to rapidly modify its structure and dynamics with respect to the evolution of external entities, identifying and solving the problems imposed by the law of requisite variety both through direct (relationship with the external actors) and indirect (definition of rules and guidelines based on the study of external social and economic environment) processes. Acting on both these processes, it is possible to obtain information needed to solve specific problems (direct adaptation) and understand rules and principles useful to solving similar problems over time (indirect adaptation).

According to Ashby's contribution, the system's ability to understand and solve a new problem is related to its capability to identify the best adaptation approach. Along such a line, complexity emerges as a consequence of the system's inability to face the problems related to the law of requisite variety choosing the *correct* process of adaptation.

Following these reflections, it emerges that the real challenge for the survival of a system is the definition of strategies and approaches inspired by the logic of adaptability in terms of a system's ability to modify its structure without revising its nature as an effect of external pressures (Alter 2008; Fricke and Schultz 2005; Zappa 1956).

By analysing the concept of adaptability in both biological and managerial domains, it is possible to state that the adaptability of a species to a specific environment can be seen as similar to the adaptation of an organisation to its context. We assume that, over time, a characteristic of the species becomes less and less appropriate, changing randomly within a certain range among future descendants. Therefore, adaptation may be understood as a suitable mechanism to conform a critical characteristic to the variations necessary to survive. From the above, it is possible to derive that the species, as a whole, learns and is influenced by a set of counter-reactions. We also have to consider the influences that the species has on its environment. Moreover, we can deduce a corollary of great importance, namely that the structure of each organic being has a close relation, which is often very difficult to see, with those of all other living beings with which it competes.

The described process does not represent a direct isomorphism for business organisations, as they do not generate descendants. However, few changes are needed to make this analogy a sufficiently coherent model. During the lifetime of an organisation, some parts of it—of its system—and its functions are replaced, and it is likely that the substitutes will be different from the originals. Undoubtedly, there will be a counterforce: adaptation continues. The real difficulty with this model is that biological species use a laborious system of random mutations to adapt their characteristics to the change that has occurred, while business organisations are unable to deal with such a random experiment. Indeed, according to Beer (1959), incorrect changes are prohibited by management since self-destruction is not allowed. This implies that, thanks to its experience, management has the ability to predict the 'behavioural model of a situation in order to choose the best path of change.

In light of these considerations, it is clear that organisations need a different way of thinking, building and developing that moves beyond the classic standards in order to survive fierce industrial competition. To help overcome the inadequacies of the traditional organisational logic, we must refer to different concepts; indeed, Stafford Beer theorised an excellent frame of reference for the study of viable systems. The model, essentially based on Ashby's law of requisite variety, assumes that 'the interests of management towards the environment are simply mediated by actual current operations it undertakes there (since in reality the operation is inserted in the environment, and the management is inserted in the transaction)' (Beer 1985, p. 34).

At this point, it should be specified that an organisation more suitable in absolute terms does not exist; being successful can be interpreted as a certain quality of the relationship between being organised and the context of reference. In other words, there is no one best way: A specific way to analyse and interpret *reality* will be valid or not depending on the context; that is, there is no right generalisation, but each model should be evaluated according to a specific context.

The result is that a modification of the context has an impact on the conditions and chance of survival; certain variations from one organisation to another, or from one variety to another, will acquire value, so the multiplication of variations within the same species increases the chances of survival in relation to other species that have not demonstrated a similar ability to adapt (Landier 1988).

The meaning for the decision makers—namely, the governing body—is that the success of management is closely related to its ability to learn from changing situations. Investing resources in the development of this perspective is unquestionably wiser than investing them in the drafting of long-run plans or in the search for universal solutions that can be valid at all times and in all places.

12.4.3 The Viable Systems Approach

An interesting contribution to a better understanding of the relevance of knowledge in the managerial field is offered by the *viable systems approach* (vSA). It is both a conceptual framework and an interpretative lens built by an Italian research group directed by Professors Gaetano Golinelli and Sergio Barile (Barile 2009; Barile and Saviano 2010, 2011; Golinelli 2010; Golinelli and Vagnani 2002; Golinelli et al. 2012). The vSA builds on the preliminary studies of Stafford Beer and on his definition of a viable system as one that is able to adapt itself, maintain an independent existence, and co-evolve with a changing environment (Beer 1972).

The principal aim of the vSA is to investigate organisations by considering their relationships with their context and the different elements that affect their decisions and processes. In doing so, a first important distinction is made between the *area of decisions* and the *area of actions*, a distinction which comes from the dichotomy between decisions and actions already defined by Beer (1972). However, the vSA enriches the dichotomy with the property of isotropy, which means that the behaviour of every system is characterised by a continuous interplay between decisions and actions. This gives the opportunity to investigate every type of organisation, group and system as well as every individual component identifying the two different areas of decisions and actions. Moreover, these two area can also be assimilated with two other areas—that is, governance and management. According to the vSA, none of these areas can be autonomous within the system, and they are in continuous interchange. In a business environment, the area of decisions needs an amplification of information variety by receiving more knowledge to operationalise decisions performed by the area of actions, while the performance of the actions of the area of decisions needs to be *compressed* to provide the essential information to the governing area to make decisions.

Thanks to the vSA, traditional boundaries dissolve and it becomes possible to focus attention on the ways in which different components of the same structure interact and a possible system oriented towards achieving a specific aim emerges (Barile et al. 2012, 2013). In such a vein, the vSA allows us to define the pathway of emergence of a system in a business environment, identifying which elements are involved in each step and how they interrelate to achieve a shared aim (Barile 2013; Saviano and Caputo 2012).

A representation of this pathway is given in Fig. 12.1, which shows the so-called *conceptual matrix* (Golinelli 2010; Golinelli and Gatti 2001). In particular, in order to understand and explain the evolution of every type of organisation, it is necessary to define the following steps:

- The conception of a *business idea*, which represents the set of basic distinctive elements which will justify competitive advantage in the market;
- The definition of a *logic structure* related to the ways to achieve the specific aim;
- The actual structure, related to the resources and the relations needed to achieve the aim;

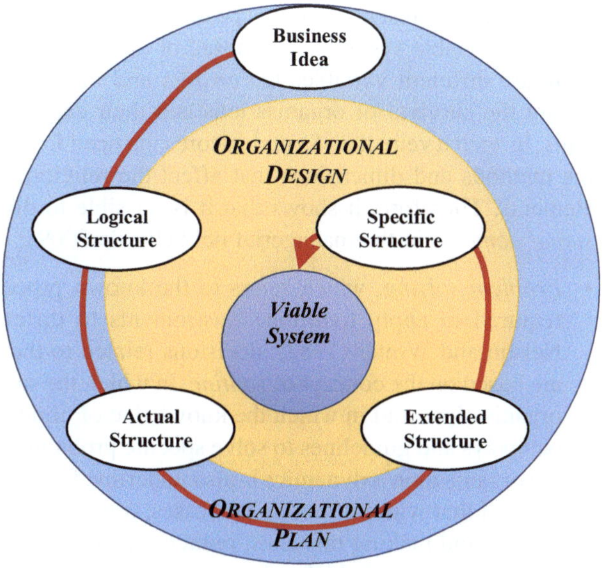

Fig. 12.1 The conceptual matrix. Source: www.asvsa.org

- The extended structure, related to the complex relationships that decision makers could activate to achieve the specific aim;
- The specific structure, related to the relationships that decision makers chose to achieve the aim.

After all these steps, the emergence of the system occurs from the actual interactions among the parties/elements that are projected into the specific structure. This is also a passage from the parties to the whole, from the abstraction to the completeness, and from the complexity to the complicatedness, since the entropy has been reduced.

This view supports the adoption of two different levels of analysis to investigate the links among the elements of an organisational configuration. In fact, from a structural point of view, the *relations* can be defined in terms of their logical or physical connection; while, from a systemic point of view, the interrelations and *interactions* can be defined in terms of the ways in which resources (tangible and intangible) are exchanged among the elements.

Finally, the conceptual matrix allows us to distinguish between two phases needed to ensure the emergence of a system related to the definition of rules and guidelines able to support the connection among the different elements in order to achieve the planned aims. In this direction, the *organisational plan* refers to the first phase of the conceptual matrix, in which the rules and pathways for the identification of the business idea and the definition of the logical and actual structure are performed; the second phase, the *organisational design,* is related to the ways in which the business idea is developed and the planned aims are to be achieved.

From this representation, it is possible to note how every type of organisation in terms of a viable system is composed of different elements and, for each of them, there are different variables to consider and include in the managerial models to ensure the survival of organisations and their capability to achieve their planned aims. In such a vein, the *vSA* underscores the need for decision makers to understand the methods and dimensions that affect the relationships among different actors/ elements. Therefore, it shows that it is possible to distinguish two different *decisional domains* in the managerial field (Barile 2009):

- *Problem solving*, which refers to the known processes and dynamics that are required to apply traditional instruments to different contexts. According to Nelson and Winter (1982), decisions related to the domain of problem solving are based on the concept of *routine*, in which the same processes are repeated by organisations and in which the knowledge of organisations acquires the form of pathways and guidelines to solve specific problems. The relevance of routines to the organisation's dynamics is also underlined by Gatti and Vagnani (2009), who showed that without routine processes, organisations risk losing control of their actions and making mistakes, reducing their probability of survival.
- *Decision making*, which refers to situations in which it is not possible to apply old instruments or approaches. These require the identification of new pathways or the development of a different perspective that is able to support the decision makers in identifying effective, efficient and sustainable solutions. In these cases, to face a specific problem, it is sometimes necessary to rethink one or more levels of the system in order to develop pathways for identifying the best solution with regard to the resources, competences and capabilities of organisations. According to Badinelli et al. (2012), decision making is based on the knowledge of organisations and requires the calculation of failures and opportunities of certain pathways selected from among the range of known opportunities.

Given this distinction, it is clear that decision making deals with chaotic and complex business environments, while problem solving can be related to complicatedness. Along such a line, the definition of effective approaches to decision making requires building instruments and pathways capable of increasing the capacity of decision makers to align their information variety to the processes and dynamics of the external context. Accordingly, the relevant lever for companies to better face social and economic challenges is the improvement of learning processes based not only on the acquisition of data related to the given social and economic dynamics, but also on the capacity to stimulate the decision makers' ability to read, analyse and explain the data on the basis of new interpretative schemes inspired by a holistic view.

12.5 Conclusions, Implications and Future Directions for Research

By analysing the changes of decision making in business management, the existence of a process of dynamic adaptation becomes clear. At the same time, both researchers and practitioners dealing with decision making have developed conceptual frameworks, instruments and approaches with the aim to adapt a system's structure and dynamics to external changes. A large segment of guidelines for the definition of efficient and effective approaches in decision-making processes have been defined as a consequence of the diffusion and understanding of specific social and economic dynamics.

Nowadays, the vibrant dynamics in which social and economic actors live daily highlights the need for a paradigmatic change in perspective towards the ways we define, imagine and live the decision-making processes. A more holistic approach is required to perceive and understand the emerging social and economic dynamics and to be able to proactively modify the structure and dynamics of companies interested in surviving over time.

According to these reflections, the *vSA* is proposed as a wide conceptual framework rooted in the systems thinking domain and able to portray in a common *picture* the most relevant variables for directly and indirectly affecting decision-making processes. Thanks to the interpretative contribution provided by the *vSA*, it is possible to emphasise that decision-making processes cannot be ex ante defined as a consequence of the application of a hierarchical approach (Caputo 2016). The building of effective decision-making processes increasingly requires the definition of participatory approaches that are able to overcome the traditional hierarchical logic in order to support the emergence of shared pathways in which all the involved actors are interested in providing individual resources to achieve both individual and collective aims (Caputo et al. 2016b; Dominici et al. 2017; Polese et al. 2016).

The underlined paradigm shift in the definition of the decision-making domain highlights several implications from a managerial and practical point of view. Specifically, from a theoretical point of view, the need arises for:

- Better investigating cognitive and psychological dimensions capable of affecting the success or failure of decision-making processes;
- Extending the domain of decision making in order to include dimensions and dynamics that are related to the external environment but are able to directly and indirectly affect the system's survival.

At same time, from a managerial point of view, the need emerges for:

- Defining new approaches to the ways in which decision-making activities are built and shared inside and outside the company.

- Improving the quality of transdisciplinary ability and the knowledge of decision makers in order to increase a company's ability to better face social and economic challenges.

Reflecting upon these implications, several future developments for research can be derived in terms of the need for defining appropriate instruments and developing qualitative and quantitative approaches useful for better investigating the analysed emerging dimensions and dynamics of decision making and their impact on a system's performance and survival.

References

Alter, S. (2008). Service system fundamentals: Work system, value chain, and life cycle. *IBM Systems Journal, 47*(1), 71–85.

Anfossi, A. (1971). *Prospettive sociologiche dell'organizzazione aziendale*. Milano: Franco Angeli.

Angaroni, G. (1977). *Elementi di economia e di gestione industriale*. Milano: ISEDI.

Ashby, W. R. (1956). *An introduction to cybernetics* (1st ed.). London, UK: Chapman and Hall.

Ashby, W. R. (1960). *Design for a brain; The origin of adaptive behaviour* (2nd ed.). New York, NY: John Wiley and Sons.

Ashley, W. C., & Morrison, J. L. (1997). Anticipatory management: Tools for better decision making. *The Futurist, 31*(5), 47–50.

Badinelli, R., Barile, S., Ng, I., Polese, F., Saviano, M., & Di Nauta, P. (2012). Viable service systems and decision making in service management. *Journal of Service Management, 23*(4), 498–526.

Barile, S. (2009). *Management Sistemico Vitale. Decidere in contesti complessi*. Torino: Giappichelli.

Barile, S. (Ed.). (2013). *Contributions to theoretical and practical advances in management. A viable systems approach (vSA)*. Roma: Aracne.

Barile, S., & Saviano, M. (2010). A new perspective of systems complexity in service science. *Impresa, Ambiente, Management, 4*(3), 375–414.

Barile, S., & Saviano, M. (2011). Foundations of systems thinking: The structure-system paradigm. In Various Authors, *Contributions to Theoretical and Practical Advances in Management. A Viable Systems Approach (vSA)* (pp. 1–24). Avellino: International Printing.

Barile, S., Pels, J., Polese, F., & Saviano, M. (2012). An introduction to the viable systems approach and its contribution to marketing. *Journal of Business Market Management, 5*(2), 54–78.

Barile, S., Carrubbo, L., Iandolo, F., & Caputo, F. (2013). From 'EGO' to 'ECO' in B2B relationships. *Journal of Business Market Management, 6*(4), 228–253.

Barile, S., Saviano, M., & Caputo, F. (2015a). How are markets changing? The emergence of consumers market systems. In G. Dominici (Ed.), *The 3rd International Symposium Advances in Business Management. "Towards Systemic Approach"* (pp. 203–207). Avellino: Busyness Systems. E-book Series.

Barile, S., Saviano, M., & Simone, C. (2015b). Service economy, knowledge, and the need for T-shaped innovators. *World Wide Web, 18*(4), 1177–1197.

Barnard, C. H. (1936). Mind in everyday affairs, a Cyrus Fogg Brackett Lecture Before the Engineering Faculty and Students of Princeton University, March 10, 1936.

Barnard, C. I. (1968). *The functions of the executive*. Cambridge: Harvard University Press.

Beer, S. (1959). *Cybernetics and management*. London: English Universities Press.

Beer, S. (1972). *Brain of the Firm, Allen Lane*. London: The Penguin Press.
Beer, S. (1985). *Diagnosing the system for organization*. New York: John Wiley.
Bohm, D. (2002). *Wholeness and the implicate order*. London: Psychology Press.
Bonazzi, G. (2002). *Come studiare le organizzazioni*. Bologna: il Mulino.
Bonazzi, G. (2007). *Storia del pensiero organizzativo*. Milano: Franco Angeli.
Boulding, K. E. (1961). *The image: Knowledge in life and society*. New York: University of Michigan Press.
Bracker, J. (1980). The historical development of the strategic management concept. *Academy of Management Review, 5*(2), 219–224.
Caputo, F. (2016). A focus on company-stakeholder relationships in the light of the stakeholder engagement framework. In D. Vrontis, Y. Weber, & E. Tsoukatos (Eds.), *Innovation, entrepreneurship and digital ecosystems* (pp. 455–470). Cyprus: EuroMed Press.
Caputo, F., Del Giudice, M., Evangelista, F., & Russo, G. (2016a). Corporate disclosure and intellectual capital. The light side of information asymmetry. *International Journal of Managerial and Financial Accounting, 8*(1), 75–96.
Caputo, F., Evangelista, F., & Russo, G. (2016b). Information Sharing and Communication Strategies: A stakeholder engagement view. In D. Vrontis, Y. Weber, & E. Tsoukatos (Eds.), *Innovation, entrepreneurship and digital ecosystems* (pp. 436–442). Cyprus: EuroMed Press.
Covey, S. R. (1992). *Principle centered leadership*. New York: Simon and Schuster.
Debeir, J. C., Deleage, J. P., Hémery, D., & Bairati, P. (1987a). Storia dell'energia. Dal fuoco al nucleare. Il Sole 24 ore, Milano.
Debeir, J.C., Deléage, J.P., & Hémery, D. (1987b). Storia dell'energia. Dal fuoco al nucleare, Il Sole 24 Ore, Milano.
Di Nauta, P., Merola, B., Caputo, F., & Evangelista, F. (2015). Reflections on the role of university to face the challenges of knowledge society for the local economic development. *Journal of Knowledge Economy, 1-19.*
Dominici, G., Yolles, M., & Caputo, F. (2017). Decoding the dynamics of value cocreation in consumer tribes. An Agency Theory approach. *Cybernetics and Systems. An International Journal, 48*(2), 84–101.
Emery, F. E. (Ed.). (1981). *Systems thinking: Selected readings*. London: Penguin books.
Formisano, V., Caputo, F., & D'amore, R. (2015). Tratti evolutivi della società della conoscenza: il contributo degli studi sulle reti nella prospettiva sistemica. *Esperienze d'impresa, 2*, 73–94.
Fricke, E., & Schulz, A. P. (2005). Design for changeability (DfC): Principles to enable changes in systems throughout their entire lifecycle. *Systems Engineering, 8*(4), 342–359.
Galbraith, J. K. (1973). *Economics and the public purpose*. London: Houghton Mifflin (T).
Galbraith, J. R. (1974). Organization design: An information processing view. *Interfaces, 4*(3), 28–36.
Galbraith, J. R. (1977). *Organization design*. London: Addison Wesley Publishing Company.
Gatti, C., & Vagnani, G. (2009). Intelligenza, compimento e vitalità nei sistemi di sistemi. *Sinergie, 80*, 59–70.
Golinelli, G. M. (2010). *Viable systems approach (VSA): Governing business dynamics*. Padova: Cedam.
Golinelli, G. M., & Gatti, M. (2001). The firm as a viable system. symphonya. *Emerging Issues in Management, 2*, 38–63.
Golinelli, G. M., & Vagnani, G. (2002). Corporate governance as a viable system: The role of intra-and inter-systemic relationships. symphonya. *Emerging Issues in Management, 2*, 54–68.
Golinelli, G. M., Barile, S., Saviano, M., & Polese, F. (2012). Perspective shifts in marketing: toward a paradigm change? *Service Science, 4*(2), 121–134.
Grönroos, C. (1994). From scientific management to service management: A management perspective for the age of service competition. *International Journal of Service Industry Management, 5*(1), 5–20.
Gross, J. J., & John, O. P. (2003). Individual differences in two emotion regulation processes: Implications for affect, relationships, and well-being. *Journal of Personality and Social Psychology, 85*(2), 348.

Kandampully, J., & Duddy, R. (1999). Competitive advantage through anticipation, innovation and relationships. *Management Decision, 37*(1), 51–56.
Katona, G. (1964). *L'"Uomo Consumatore"*. Milano: Etas Kompass.
Landier, H. (1988). *L'impresa policellulare. Per un management del disordine*. Milano: Guerrini e Associati.
Lengnick-Hall, M., & Lengnick-Hall, C. (2002). *Human resource management in the knowledge economy: New challenges, new roles, new capabilities*. New York: Berrett-Koehler Publishers.
Maier, C. S. (1970). Between Taylorism and technocracy: European ideologies and the vision of industrial productivity in the 1920s. *Journal of Contemporary History, 5*(2), 27–61.
Maraschini, F. (2002). *La teoria dell'organizzazione come teoria della conoscenza* (Università degli Studi di Udine, Working Paper, MOS. 01-02).
Marcel, A. J. (1983). Conscious and unconscious perception: An approach to the relations between phenomenal experience and perceptual processes. *Cognitive Psychology, 15*(2), 238–300.
Maslow, A. H. (1975). *Motivation and personality*. New York: Harper & Row.
Mella, P. (2005). *Dai sistemi al pensiero sistemico*. Milano: Franco Angeli.
Merkle, J. A. (1980). *Management and ideology: The legacy of the international scientific management movement*. Berkley: University of California Press.
Minsky, M. (1988). *Society of mind*. New York: Simon and Schuster.
Momigliano, F. (1975). *Economia industriale e teoria dell'impresa*. Bologna: il Mulino.
Mooney, J. D. (1947). *The principles of organization*. New York: Harper.
Nelson, R. R., & Winter, S. G. (1982). The Schumpeterian tradeoff revisited. *The American Economic Review, 72*(1), 114–132.
Oliver, C., & Holzinger, I. (2008). The effectiveness of strategic political management: A dynamic capabilities framework. *Academy of Management Review, 33*(2), 496–520.
Pascale, T. R. (1999). Surfing the edge of chaos. *Sloan Management Review, 40*(3), 83.
Polese, F., Caputo, F., Carrubbo, L., & Sarno, D. (2016). The value (co)creation as peak of social pyramid. In T. Russo-Spena & C. Mele (Eds.), *Proceedings 26th Annual RESER Conference, "What's ahead in service research: New perspectives for business and society"* (pp. 1232–1248). RESER, University of Naples "Federico II".
Ramage, M., & Shipp, K. (2009). *Systems thinkers*. New York: Springer Science & Business Media.
Saviano, M., Caputo, F., Formisano, V., & Walletzký L. (2016). From theory to practice in systems studies: A focus on Smart Cities. In D. Vrontis, Y. Weber, & E. Tsoukatos (Eds.), *Innovation, entrepreneurship and digital ecosystems* (pp. 443–454). Cyprus: EuroMed Press.
Senge, P. M., & Sterman, J. D. (1992). Systems thinking and organizational learning: Acting locally and thinking globally in the organization of the future. *European Journal of Operational Research, 59*(1), 137–150.
Sombart, W. (1969). *Il capitalismo moderno*. Torino: Utet.
Stalk, G., Evans, P., & Shulman, L. E. (1991). Competing on capabilities: The new rules of corporate strategy. *Harvard Business Review, 70*(2), 57–69.
Stenning, K., & Lambalgen, M. (2005). Semantic interpretation as computation in nonmonotonic logic: The real meaning of the suppression task. *Cognitive Science, 29*(6), 919–960.
Sterman, J. D. (2000). *Business dynamics: Systems thinking and modeling for a complex world*. Boston: McGraw-Hill.
Taylor, F. W. (1914). *The principles of scientific management*. New York: Harper.
Taylor, F. W. (2004). *Scientific management*. London: Routledge.
Thompson, J. D. (1967). *Organizations in action: Social science bases of administrative theory*. New York: Transaction Publishers.
Von Bertalanffy, L. (1969). *General system theory: Foundations, development, applications*. New York: George Braziller.
Winsor, R. D. (1992). Talking the Post-Fordist Talk,... but walking the post-industrial walk. *Journal of Organizational Change Management, 5*(2), 61–69.
Zappa, G. (1956). *Le produzioni nell'economia delle imprese* (Vol. 1). Milano: Giuffrè.

Chapter 13
An Overview of the Contribution of Systems Thinking Within Management and Marketing

Roberto Bruni, Luca Carrubbo, Ylenia Cavacece, and Debora Sarno

Abstract In recent years, common interests have led the fields of marketing and management to differentiate themselves, even in similar research areas. This chapter provides an overview of systems thinking in these two disciplines, reading the main concepts in an integrated way according to relational and value perspectives. A systematic literature review is conducted using the Web of Science, Scopus and Google Scholar databases, focusing on applications of the Viable System Model and the viable Systems approach. Due to the adoption of systems thinking, the paper can serve to spur further studies to better define the boundaries and the eventual inclusion of marketing in management and vice-versa.

Keywords Systems thinking • Systems theories • Viable systems approach • Viable system model

13.1 Introduction

Systems thinking addresses the process of thinking by using *system* ideas in natural, designed or management systems, while *systems theory* describes the theory of systems, taking as given the status of systems as a thing in the world (Checkland 1999). Drawing on Von Bertalanffy's (1956) seminal work, Maturana and Varela and

---

Although the chapter is the result of the joint work of the four authors, paragraphs 1 and 4.4 are attributed to Luca Carrubbo; paragraphs 2 and 6 are attributed to Debora Sarno; paragraphs 3 and 5 are attributed to Roberto Bruni; paragraphs 4.1, 4.2 and 4.3 are attributed to Ylenia Cavacece.

R. Bruni (✉)
Department of Economics and Law, University of Cassino and Southern Lazio, Cassino, Italy
e-mail: r.bruni@unicas.it

L. Carrubbo • Y. Cavacece
Department of Business Science, Management and Innovation Systems, University of Salerno, Salerno, Italy
e-mail: lcarrubbo@unisa.it; ycavacece@unisa.it

D. Sarno
Department of Economics, University of Foggia, Foggia, Italy
e-mail: debora.sarno@unifg.it

S. Barile et al. (eds.), *Social Dynamics in a Systems Perspective*, New Economic Windows, DOI 10.1007/978-3-319-61967-5_13

Senge shed light on numerous phenomena (Maturana and Varela 1975; Senge and Sterman 1992) by focusing on the relationships between different and sometimes interacting elements. Their contributions arose from the need to better understand the whole—*shifting the focus from parts to the whole* (Checkland 1997; Weinberg 2001; Jackson 2003)—and they investigated the dynamics of relationships by seeking to better define medium and long-term interactions.

This chapter provides an overview of systems thinking in business studies, with an emphasis on marketing and management. Although scholars and practitioners (Birnbaum et al. 1990; Barile et al. 2012a) have addressed the relevance of management's role in business studies with a focus on marketing, there is also a lively international debate that addresses the relationship between marketing and management. This work approaches the main concepts of marketing and management in an integrated way, and it avoids giving preference to one issue over the other to overcome a reductionist vision of marketing, which has *generally been considered to be just a useful communication tool.* The current paper aims to identify the common elements and paths of the relationships and value at the root of business management. Further, management can be considered to be a set of rules and models that are aimed at efficiency, organization, and strategic-thinking.

The chapter is structured as follows: Sect. 13.2 provides a brief literature review of different perspectives on systems theories; Sect. 13.3 presents the methodology that is used to investigate marketing and management according to systems thinking through a systematic literature analysis on the topics; Sect. 13.4 shows the findings of the literature review and explores the main traits of systems thinking, the Viable System Model and the viable Systems approach (Sect. 13.4.1) and it analyses the contribution of systems thinking to marketing and management (Sects. 13.4.2 and 13.4.3), and, finally, it compares the structural, behavioural and systems perspectives and their contribution to value creation (Sect. 13.4.4). Discussion (Sect. 13.5) and Conclusions (Sect. 13.6) follow.

13.2 Systems Theories Perspectives

A *system* can be defined as an entity that is made up of parts (elements) that are connected in a mutual, interactive relationship and share a specific aim (von Bertalanffy 1962; Parsons 1965; Barile 2006, 2008). The contexts and situations of the interactions and relationships of different entities such as nature, science, and society can be investigated as systems (Tien and Berg 2003).

Von Bertalanffy (1956) was among the first scientists to consider a system as a scientific paradigm and to study the underlying relevance of interactions among systems. *Systems theories* are useful to identify general references that can be extended to every type of relational activity between actors and/or elements, which can support the understanding of complex phenomena.

According to Mele et al. (2010), several perspectives have paved the way for the development of systems theories including General Systems Theory (GST),

Cybernetics (Beer 1975), Organization, Biology (Maturana and Varela 1975) and Sociology. Indeed, different approaches have been dedicated to systems analysis including viable systems (Espejo and Harnden 1989; Barile 2009), service systems, system dynamics and smart systems (Barile and Polese 2010a, b; Demirkan et al. 2011a, b), intelligent systems, economics systems, eco-systems (Wieland et al. 2012; Vargo and Lusch 2016; Polese and Carrubbo 2016) and many others.

Within the *GST*, two relevant approaches have assumed a growing importance and provide a strong contribution to the understanding of the relationships and interactions inside and between systems and the environment: the Viable System Model (VSM) and the viable Systems approach (*vSa*). According to these theories, the adaptability of systems to environmental changes is of great importance and is related to the informative variety (Ashby 1958) and systems openness (Katz and Kahn 1978), which contributes to specific actions such as organization and self-organization. The *Viable System Model (VSM)* (Beer 1972; Espejo 1999) is rooted in cybernetics and describes a system as an entity that is adaptable for the purpose of surviving in its changing environment (Beer 1972). The model focuses on conceptual tools to obtain an understanding of the organization of systems to redesign them through: (i) changes in management; (ii) understanding the organization as an integrated whole; and (iii) evaluating the essential functions of implementation, coordination, control, intelligence and policy (Beer 1972; Espejo and Harnden 1989; Espejo 1999; Christopher 2007). According to Stafford Beer (1975), a viable system is able to survive and remain closed and integral; it is homeostatically balanced both internally and externally, and it has mechanisms and opportunities to expand and learn, to develop and adapt, in other words, to became even more effective in its environment.

The *viable Systems approach (vSa)* (Golinelli 2005, 2010; Barile 2006, 2008) identifies certain relevant differences, for example, related to the new interpretation of consolidated strategic organizational and managerial models such as sub-systems and supra-systems. Sub-systems focus on the analysis of the relationships among enterprises' internal components, while supra-systems focus on the connections between enterprises and other influencing systemic entities within their context (Golinelli 2000, 2005; Barile 2006, 2008). Other main differences can be identified from the emergence perspective as opposed to the functionalist perspective, and from the constructivist approach as opposed to the cybernetic approach; network logic, which overcomes the hierarchical model, can represent the relationships between actors (Barile et al. 2015).

Open systems are of great importance in *business organization studies* (Katz and Kahn 1966), with a focus on social systems and the relevance of energy-material exchanges between organizations and environment. In this context, Emery and Trist's (1960) contribution underscores two company components that are considered to be a system: the social component (people) and the technical component (technology and machines).

Biology and Sociology are the two main disciplines that focus on systems theory. Based on the work of Maturana and Varela (1975), the relevance of system adaptability has been recognized as arising from the internal ability to adapt to emerging changes. Their interpretation of *autopoiesis—the capability to reproduce and maintain itself*—has also been widely used in Sociology (Luhmann 1990), in which a system is defined as an area in which the complexity can be reduced (the internal side of the system). This approach contributes to a better understanding of using self-regulation and self-organization to manage complexity. According to the previous findings, systems can be considered learning entities that are able to understand the surrounding environment due to knowledge functions that contribute to the reduction of internal entropy and to the increase of external entropy (Von Foerster 1981).

With regard to *System Dynamics and Smart Systems*, many past and current studies have contributed to a better understanding of the influence of systems theories on business topics. The most relevant concepts for these two complementary approaches are learning relevance, self-regulation, and reconfiguration (Sterman 1994); intelligence is considered to be both internal and external to the system. In particular, in terms of "smart" systems (Basole and Rouse 2008; Demirkan et al. 2008), the *vSa* offers a useful contribution (Barile and Polese 2010b; Polese et al. 2012). In these studies, systems are considered to be smart, searching for dynamic and intelligent IT-based services, and they are characterized by viable behaviour able to promote long-lasting system competitiveness and performance (Napoletano and Carrubbo 2011).

This brief analysis underscores the relevance of the elements and characteristics that have been derived from systems theories, mainly the role of relationships, adjustments, environment, and complexity.

13.3 Methodology

This study aims to provide an overview of the meanings and approaches that are related to management and marketing in business studies. It is based on a literature review of the contributions that analyse management and marketing through the lens of systems thinking theories. The review method is conducted in three consecutive stages:

1. Literature search;
2. Assessment and clustering of the evidence base;
3. Analysis and synthesis of the findings.

Literature Search We opted to search the relevant literature in the electronic databases Scopus, Web of Science and Google Scholar, which contain extensive literature on the topics of interest. We considered the time range from 2000 to 2016 to identify the most recent evolutionary trajectories of the debate on management and marketing meanings. The searches were performed using keyword combinations, looking for matches in the fields (Title) and/or (Abstract) and/or (Article

Keywords). We employed two keyword combinations and located a total of 332 results:

- "Systems thinking" AND "Management";
- "Systems thinking" AND "Marketing".

Assessment and Clustering of the Evidence Base The first examination of the identified publications made it clear that there were redundant entries that were not related to this particular study. Hence, the gathered evidence base was examined to determine the studies most relevant to the particular focus and scope to be applied here. For example, searching with the keyword combination ("systems thinking" and "management") returned a publication titled "An essential distinction of agile software development processes based on systems thinking in software engineering management" (Wendorff 2002), which has a different focus than that of this study. Each title and abstract and, in case of doubts, the introduction and conclusion sections of the paper, were carefully read; sometimes the entire article was skimmed before the final decision was made.

In particular, the research on management identified 320 results:

- 122 were discarded (being not relevant for the research criteria) after reading the title
- 60 were discarded (being not relevant for the research criteria) after reading the abstract
- 9 were discarded (being not relevant for the research criteria) after reading the text

Then, 129 were saved and analysed.

The research on marketing identified 12 results:

- 3 were discarded (being not relevant for the research criteria) after reading the title
- 9 were added from the bibliographies of the analysed papers, being considered interesting and useful for the research

In total, 18 were analysed.

Ultimately, 147 articles were selected for the final analysis.

Some of the selected articles apply *systems thinking* in general to management and marketing, while others focus more on specific theories such as *VSM* or *vSa*; to facilitate the analysis and synthesis of the results, the contributions that belong to the two main categories were grouped, and a third cluster was left for other approaches.

We noted that only a few works concerning management addressed management science as a whole, while the majority were focused on complexity theory and other specific areas of operational management, which, for a faster and better analysis, we divided into different thematic groups: knowledge, project, supply chain, innovation, quality, risk and healthcare management and decision making.

Analysis and Synthesis of the Findings After the analysis, we explained the main findings related to the contribution of the systems theories in understanding management and marketing and identifying the different perspectives that emerge in the literature.

13.4 Literature Review on Management and Marketing in Systems Thinking

13.4.1 The Main Traits of Systems Thinking, VSM and vSa

The analysis of the selected papers shows that most of the contributions use *systems thinking* as a framework that can be applied to the study of phenomena. Several systems theories have also been applied to management and marketing studies; however, the most common one seems to be complexity theory. There is a widely held view that systems thinking is superior to other approaches in addressing complexity (Maani and Maharaj 2004); thus, it considers strategy in terms of 'order out of chaos' and defines strategic decision-making as a complex activity because it involves not only different issues but also many interacting factors and stakeholders (Sheffield et al. 2012; Donald 2010; Powell 2004). Several authors have drawn on systems thinking and complexity theory to re-conceptualize and manage organizations as complex adaptive systems (Reiman et al. 2015; Rabaey 2013).

Another widespread theory in management studies is *VSM*, which has been considered to be a theoretical framework that leads to a better understanding of *sustainability's* role in several areas, such as:

1. Complex contexts (Espinosa et al. 2008);
2. Knowledge management (Choi and Hilton 2005; Ganzert et al. 2012; Paucar-Caceres and Pagano 2009);
3. Visualization of viable inter-organizational relationships that can be integrated along a whole supply chain and in product development (Puche et al. 2016; Chroneer and Mirijamdotter 2009);
4. Adaptability, in terms of the challenges that must be faced to link changes among the stakeholders (Murad and Cavana 2012).

However, *vSa* is more prevalent in marketing studies, and it is considered to be a useful approach to understand not only the dynamic interactions in many-to-many marketing networks (Barile and Polese 2010a, b) but also the role of customers in value co-creation processes (Barile et al. 2012a, b).

Focusing on systems thinking, it emerges that *vSa* has been widely applied to management studies to explain managerial theory through systemic paradigms. This process is closely connected to the operational research, the study of internal processes, planning and control, problem solving and decision making, which can be applied to specific areas such as knowledge, project risk, innovation and supply chain management, and quality control (Carrubbo et al. 2015; De Santo et al. 2011). However, few studies have combined the two concepts.

In marketing, systems thinking has been mainly applied to two macro-areas: the study of networks and value co-creation, and social marketing. Therefore, these studies are focused on relationships and interactions between all actors, while the investigation is gradually shifting to the external dimension of organizations, especially with regard to customers.

13.4.2 Systems Thinking and Management

In management studies, the role of systems thinking involves the following activities (Mingers and White 2010; Allen 2000; Polese et al. 2010; Barile et al. 2014):

- Holistic observation of phenomena, which is considered to be a set of different elements that interact within a specific environment;
- Attribution of a growing importance to relationships or interactions between elements, which is considered to be more significant than the elements themselves in defining a system's behaviour;
- Definition of systems' hierarchical levels of the properties that emerge at different levels and of the mutual causality within and between levels.

Systems thinking has been deeply connected to the development of *operational research and management science* (Mingers and White 2010; Jackson 2009; Hitchins 2003) and change management processes (Haines 2000) with a focus on the internal managerial cycle that includes planning, organizing, leading, and controlling paradigms. In the same vein, Allen (2000) sees the organization as a system and the scheduling of activities as dependent on different organizational levels.

Systems thinking has been largely applied to *project management* as a flexible approach to the management of innovations (Cavaleri et al. 2012), complexity, and uncertainty to make innovation projects as successful as possible (Bendoly 2013; Williams 2016; Chronéer and Backlund 2015; Costello et al. 2002; Lyneis and Ford 2007; Mawby and Stupples 2002; Winter and Checkland 2003).

However, in *knowledge management*, a systems approach fosters the internal dialogue and the resource exchange (Mele et al. 2010; Rana et al. 2013) that support the ability to generate new ideas, processes, and products (Urze and Abreu 2014). This approach also enables the achievement of a complete view of the whole organization (Ndlela 2014), which allows for the identification of benefits and opportunities (Barcelo-Valenzuela et al. 2008). In this way, managers can make better and informed decisions in terms of asset allocation and management (Sole and Schiuma 2010).

In recent years, several researchers have shown that systems thinking can be fundamental to the analysis of complex organizational operations, for example, those that are related to supply chain management (Beth et al. 2003; Holweg and Pil 2008; Moon and Kim 2005). According to Maull et al. (2012), the use of systems thinking can lead the analysis to overcome boundaries, hierarchies, and mechanisms of control in the supply chain.

Finally, other managerial processes have been extensively analysed according to systems thinking, including risk management (O'Donnell 2005; Kamppinen, et al. 2008; Lee and Green 2015), innovation management (Shen et al. 2009; Kong and Li 2007, 2008; Xiang-yu and Xiang-yang 2007), and quality management (Wolf et al. 2011; Chen et al. 2014; Guerreiro et al. 2014; Conti 2010).

Moreover, the application of systems thinking has been extended beyond the traditional operational boundaries, as confirmed by its application to specific areas such as healthcare. In fact, most of the actual healthcare research stresses the systems approach to assume a holistic view of a "system" at every operational level and to involve a growing number of stakeholders, particularly patients (Mutingi and Mbohwa 2014; Pentland et al. 2014; Waliullah and Schell 2013; Paina et al. 2014; Turnbull 2002; Karppinen et al. 2014; Adam 2014).

13.4.3 Systems Thinking and Marketing

In marketing, systems thinking has been mainly applied to the study of relationships among the actors (including customers) who are involved in value creation and delivery. Therefore, the focus has shifted from internal processes and operations to interactions with the external environment. One of the most investigated marketing frameworks according to systems thinking is *network theory*. In particular, a marketing system is defined as a network of individuals, groups, and/or entities that are linked directly or indirectly through sequential or shared participations in an economic exchange (Layton 2007). This network can also be considered to be the sum of the patterns that emerge from transaction flows (Layton 2011). Networks are not merely networks (aggregations of relationships); they are considered to be dynamic (Vargo and Lusch 2011) and open systems that are able to improve not only system sharing or the application of resources but also their own state, gaining external resources (Spohrer et al. 2008). *VSa* supports the understanding of dynamic interactions in many-to-many marketing networks (Barile and Polese 2010a, b).

Another marketing framework that has been analysed according to systems theories is that of *value co-creation*, which is deeply related to Service Research. As Barile and Polese (2011) argued, given the systemic nature of value creation, managers must adopt a systemic approach, which leads to a general observation of complex phenomena that facilitate value exchanges with customers (Vargo et al. 2008; Lusch et al. 2009). Moreover, value can be specifically accessed on a relative basis, in other words, through a comparison with competitors' offerings. Consequently, according to Barile et al. (2012a, b), managers should adopt a systemic approach that is rooted in a wider perspective that includes customers, partners, competitors, and other actors.

The contribution of *vSa* to marketing theories mainly comes from its wider systemic perspective, and it suggests direct efforts to offer increased dignity to the marketing debate; thus, more general theories facilitate the understanding of major changes in market conditions and the usefulness of technological advance (Barile et al. 2012a, b). In highly competitive contexts, the growth is rapid, the innovation is rich, the local conditions are idiosyncratic, and the technological options are increasingly complex. Consequently, marketing managers must understand the dynamics so they can affect industry structure. This allows them to assess market strategic value, adopt a system methodology, and develop a holistic frame of reference that can ultimately allow them to focus on relevant issues and avoid the endless search for more details and the proliferation of useless information (Pagani and Otto 2013).

A systems theory approach, which considers society to be a *complex adaptive system*, is suggested as a useful framework for social marketing campaigns because it can support new identities and increase sustainable behavioural changes (Conroy and Allen 2010). It has also been reported that all business exchanges involve systems and are characterized by a high degree of complexity, which is higher than is apparent. Moreover, systems thinking helps decision makers to more deeply understand the organizational problems that they face (Woodside 2006).

13.4.4 The Structural, Behavioural and System Perspective in Systems Thinking and Implications for the Value Creation Process

According to systems thinking and its paradigmatic developments, reflections about complexity and its management, as well as the interpretative consequences of analysed phenomena, are fundamental. In fact, phenomena can also be analysed assuming a structural and behavioural perspective, which can provide different insights into the way in which new systems aimed at value creation emerge and how they might be managed (Barile et al. 2013b; Carrubbo 2013).

No system is equal to another; however, each is characterized by specific "structural" elements that led to its own creation and the knowledge (technicalities), practical experience (practicalities), and competences (skills and abilities) that arise over time and as the result of interactions (even when they are not conscious).

The understanding of the leverages that can be used to promote the development and the implementation of a synergy among systems (intended as entities) is fundamental. This can be useful in several contexts, such as teaching (in a university classroom), working (during a programming meeting), security (during "truth" talks or questioning), in the social professions (psychoanalytical treatments), and in economy and business areas by considering the potential interactions between systemic-viable entities that act in a similar context. In all of these cases, the investigation of the behaviours and reactions that emerge from a direct comparison of subjects is fundamental to classify the emerging results, create statistics, and build forecasting or interpretative models that can be used in heterogeneous applicative contexts.

However, according to systems thinking, the managerial perspective can change dramatically because the interpretative tools become merely a part of an identitarian path that is aimed at structuring formalizations, languages, and theoretical purposes. A schematic synthesis of this paradigm's ability to simplify and summarize (which Barile and Golinelli have developed and advanced in recent years) has led to the emergence of structural, behavioural, and systemic-viable considerations that have the capacity to identify the main features that characterize a system over time, as is shown in Table 13.1.

Following the above-listed definitions of the main founders of this conceptual movement, it is possible to achieve a graphic representation of a viable system's building and functioning and its recurring internal scheme and/or structure.

According to the new systemic paradigm, Fritjof Capra (2002) argued that the relationship between the parts and the whole is inverted. The properties of the parts can be understood only *in light of the dynamics as they are related to the whole. At last, the parties do not exist.* What we call part is merely a configuration in a close network of relationships.

Due to the organization of internal components and the activation of integrative resources, value co-creation can occur, which enhances systems competitiveness and, consequently, improves their ability to survive. These "relational" systems are open systems that are embedded in the context in which they act and from which they can gain the external resources that are needed for their inner objectives of development and achievement. One of the most important inferences of this purpose is the rationalization and the subsequent management of decision-making, which is aimed at designing personal cognitive alignment according to a value perspective. If a satisfying decision arises from knowledge and informative resonance between the involved decision makers, the reciprocal value system must be tested to better understand the element that influences the complexity management. The client focus is also emphasized due to historical suggestions that were offered by Customer Relationship Management (CSR), Total Quality Management (TQM) or the Total Relationship Marketing, as being coherent with a competitive and sustainable approach to relationship management, which is fundamental in terms of value creation.

Finally, according to *vSa*, value creation processes synthetize a firm's ability to develop a coherent level of "consonance" and "resonance" in its own context. This context is characterized by several sub-systems, and the retention of productive resources is more or less critical in terms of co-creation optimization. In this process the implementation and the support of strategic control are fundamental not only to describe and monitor organizational processes but also to shift the mission and strategies in an optimized performance that is aimed at value creation and sharing.

Table 13.1 Different perspectives in systems thinking

Object of analysis	From a structural perspective	From a behavioural perspective	From a systems perspective
Origin and scope	**The structure of an organization originates from a given set of shared rules and relational connections** The structure aims to survive in function of various systems associated with it, even if not at the same time for each of them	Visible skills are a result of the **inherent capabilities**	A viable system lives and its goal is to **survive** in an environment populated by other viable systems The **viability** is commensurate with the realization of the dispositions to change
Relations and Interactions	**The system 'realizes' the structure and the relationship qualifies both of them**	**From the same relation originate more interactions**, respecting the same distinction between function and role in the moment in which the second can express the first Education is the form; the routine determines that the act is substance	Each **context is subjectively defined** and extracted by a general environment by each viable system's decision-making body, and in it the system is immersed adaptively
Perception of the context and adaptability	The **contexts are subjective** as a function of specific objectives and changing	The **rule is the application of a law**, and determines how often the rule itself can also change, if considered to be positive	Each viable system distinguishes and identifies the various supra-systems **relevant in its context**, because of its specific end goal A viable system has the ability to regulate and manage **independently** the dynamics of its **adjustment**
Preconditions to take actions	**Contingency** is influence, **planning** is critical, their composition is relevant	The **categorical values** form the basis for a personal interpretation of the events	The **convergence of the systems** and their entities of reference towards the same point starting from different initial positions is defined consonance (synthesis of compatibility, tune, affinity, etc.) and the variation of the gradient of this displacement vector (with a defined direction and verse) represents its measurement

(continued)

Table 13.1 (continued)

Object of analysis	From a structural perspective	From a behavioural perspective	From a systems perspective
Acting	The **supra-systems are capable of influencing the decisions of a system,** in particular in direct effects on its own sub-systems	The interpretation schemes shape the **information** and categories establish their **priorities** The **choices** represent the realization of the decisions as a result of the action of the interpretation scheme	The acceleration (wanted) of this relation between (among) two (or more) elements is **resonance** (defined as the modification of consonance over time)

Elaboration from: Barile et al. (2013a)

13.5 Discussion

In recent years, despite their pertinence and specialization, marketing and management have been considered to be either synonymous or different; however, somewhat similar schemes and thinking paradigms exist, or they can be seen as integrated tools that can contribute to the definition of strategy and building the path towards gaining competitive advantage with a focus on value and relationships. These fundamental concepts fit with systems thinking, which is sensitive to the management of value and relationships—*especially in the direct application to the companies*—and to the dynamic approach to the study of interactions between parts with the aim of benefiting a whole system.

The interaction and its value are considered to be the cause and effect of the value that derives from systemic relationships between elements of systems that are characterized by involvement, knowledge—*in value recognition*—and the ability to act. These are properties of systems (individuals, groups of functions) that belong to companies, which are called upon to manage complexity in marketing and management activities and are required to give significance to the relationships and their achievement as well as to explain the value creation process, managing the relationship between the context and the economic environment. In this sense, marketing and management could be supported by system theories in every type of organization, in particular in firms that must manage complexity and sustainability with regard to the environment.

The literature review underscores that systems thinking has been widely applied to management, while few studies have focused on marketing and consequently, these disciplines have sometimes been investigated in different ways. In general, different key concepts emerge in management: complexity such as the variety and variability of the external and internal phenomena that affect plans, forecasts, company performance and strategy as well as studies on decision making and problem solving. The relevant systemic concepts that are applied to the management research are holism, autopoiesis, homeostasis, and layer hierarchy, while the most discussed systems theories are the complexity theory, cybernetics, system dynamics, Soft OR, VSM, and critical systems. However, in marketing, systems thinking is mainly applied to two macro-areas: network studies and value co-creation and social marketing. Therefore, the focus is on the relationships and interactions among actors both internally and externally to the organization.

The most relevant systemic properties are system openness, resource exchange with the external environment and, in this case, dynamic system vision. The need for the management of this growing complexity also affects marketing. In this case, it is related to market dynamism, consumer attitude and relationship unpredictability.

13.6 Conclusions and Managerial Implications

This research shows the orientation of the use of systems thinking among researchers and business studies with a specific focus on marketing and management. It should also be noted that, in general, it is not simple to distinguish between marketing and management studies when they address certain common topics. However, according to both management and marketing logic, the approach is often traditional, and if the marketing interpretation is mainly closer to sales and communications, the management interpretation is closer to activities and operation, even if some of the contributions are focused on business governance.

This study underscores the need to overcome the traditional logics that practitioners often assume to support the shift towards relational concepts and value, which are considered to be the base concepts to stimulate a firm's value proposition. In the international context, the gap between marketing and management is narrowing and, consequently, systems theories may be considered to be specific frameworks that can be used to pave the way for meanings, rules, and models integration; in several cases, the need for interpreting these two logics also stems from systems theories. This situation seems to be due to specific schools of thought and practices, such as the traditional logic that considers marketing to be a "function"—*and for this reason marketing is included in management, thus systems thinking may contribute to a better understanding of firm dynamics*—and the more advanced logic that considers marketing as a strategic system of activities and tools to manage and stimulate the emersion of value and relationships that work for the competitive advantage of companies. According to these advanced thoughts, the sensitivity to market dynamics and to internal changes represents some of the most important managerial competence (top management) that should be developed in a "marketing and systemic logic". Consequently, it is possible to consider that the differences between marketing and management are becoming smoother.

In light of the reported experiences and future research trends, this chapter argues that there has been a great deal of conceptual research, even if the literature still calls for further applied study. In particular, among business studies, systems theories are certainly important and can be considered to be an interesting and broad contribution to the comprehension of the nature and dynamics at the root of relationships and interactions between the elements that allow companies to be considered systems embedded in systems.

The research evolution has led to knowledge advancements, and systems studies are based on the integration of different paradigms that use different methods and approaches to better understand the relational dynamics among interacting elements. The "arising" system is one of the most representative statements in the research on systemic dynamics, enhancing the comprehension of the process of definition of the common core of systems and, consequently, whether they are working (activating themselves) to reach a shared goal.

The marketing and management tradition reveals some important corporate needs including the definition of a target market that is able to capture a latent demand and the organizational need to adapt to a changing environment. *De facto*, the latter is a very complex feature, which has been applied over the past years to many different fields and, above all, has been able to stimulate several researchers to define the optimal complexity model. It is evident that companies are not only made up of people, but that they also have the skills and competencies that allow them to gradually adjust, transform, and restructure themselves through system interactions (Golinelli 2005).

Whereas the different approaches to system analysis are characterized by some common elements and, above all, are based on recurring and fundamental concepts such as relationships, value, complexity, and adaptability, it may be appropriate to bring marketing and management research to higher levels, eventually reducing the differences among their approaches. However, the literature is still lacking a shared definition of the edges between the two research areas; thus, when this difference arises, marketing tends to be interpreted as a set of techniques based on communications, sales and, in some cases, on market research or models that are used to identify investment optimization and to reach specific targets. The same holds for management even if, according to systems logic, there is always the need to root managerial studies in cost optimization, sometimes leaving aside all of the value components that are different from profit. A systemic approach might broaden managers' horizons; thus, a wider vision of the roles of relationships and value as well as a less narrow approach to profit or efficiency might offer more opportunities to provide value.

Systems thinking might also contribute to the definition of new approaches to marketing and management that could be applied in specific research areas and considered as a comprehensive approach to scientific research. Systems theories' contributions might especially arise from their systemic, inclusive, relational nature, and, above all, from the importance that is given to the relationships with the surrounding environment. Therefore, even if practitioners and scholars have still not defined the edges of marketing and management, the theoretical frameworks of systems theory might contribute to the definition of a possible vision that, before their nature and function are defined in the near future, might express the necessity for both the research areas to base their investigation in relationships and value. The eventual differences will be identified according to a specific approach to the starting elements of relationships and value.

References

Adam, T. (2014). Advancing the application of systems thinking in health. *Health Research Policy and Systems, 12*(1), 50.

Allen, W. (2000). The basic management cycle: A systems approach to the management process. In *Proceedings of ICSTM 2000: International Conference on Systems Thinking in Management* (pp. 39–43), Geelong, 8–10 November.

Ashby, R. W. (1958). General systems theory as a new discipline. *General Systems, 3*, 16.

Barcelo-Valenzuela, M., Sanchez-Schmitz, G., Perez-Soltero, A., Martin-Rubio, F., & Palma, J. (2008). Defining the problem: Key element for the success of knowledge management. *Journal of Knowledge Management Research and Practice, 6*(4), 322–333.

Barile, S. (2006). *L'impresa come sistema. Contributi sull'Approccio Sistemico Vitale* (I ed.). Torino: Giappichelli.

Barile, S. (2008). *L'impresa come sistema – Contributi sull'Approccio Sistemico Vitale* (II ed.). Torino: Giappichelli.

Barile, S. (2009). *Management Sistemico Vitale*. Torino: Giappichelli.

Barile, S., & Polese, F. (2010a). Linking viable systems approach and many-to-many network approach to service-dominant logic and service science. *International Journal of Quality and Service Sciences, 2*(1), 23–42.

Barile, S., & Polese, F. (2010b). Smart service systems and viable service systems: Applying systems theory to service science. *Journal of Service Science, 2*(1/2), 20–39.

Barile, S., & Polese, F. (2011). The viable systems approach and its potential contribution to marketing theory. In Various Authors (Ed.), *Contributions to theoretical and practical evidences in management. A viable systems approach (VSA), ASVSA – Association for research on Viable Systems* (pp. 139–173). Avellino: International Printing.

Barile, S., Pels, J., Polese, F., & Saviano, M. (2012a). An introduction on the viable systems approach and its contribution to marketing. *Journal of Business Market Management, 2*(54–78).

Barile, S., Saviano, M., Polese, F., & Di Nauta, P. (2012b). Reflections on service systems boundaries: A viable systems perspective. The case of the London Borough of Sutton. *European Journal of Management, 30*(5), 451–465.

Barile, S., Carrubbo, L., Iandolo, F., & Caputo, F. (2013a). From 'EGO' to 'ECO' in B2B relationships. *Journal of Business Market and Management, 6*(4), 228–253.

Barile, S., Polese, F., Calabrese, M., Iandolo, F., & Carrubbo, L. (2013b). A theoretical framework for measuring value creation based on viable systems approach (VSA). In S. Barile (Ed.), *Contributions to theoretical and practical advances in management viable systems approach*. Rome: ARACNE.

Barile, S., Polese, F., Saviano, M., Pels, J., & Carrubbo, L. (2014). The contribution of vsa and sdl perspectives to strategic thinking in emerging economies. *Managing Service Quality, 24*(6), 565–591.

Barile, S., Sancetta, G., & Saviano, M. (2015). *Management. Il modello sistemico e le decisioni manageriali* (Vol. 1). Torino: G Giappichelli Editore.

Basole, R. C., & Rouse, W. B. (2008). Complexity of service value networks: Conceptualization and empirical investigation. *IBM Systems Journal, 47*(1), 53–70.

Beer, S. (1972). *Brain of the firm*. London: The Penguin Press.

Beer, S. (1975). *Platform for change*. London: John Wiley.

Bendoly, E. (2013). System dynamics understanding in projects: Information sharing, psychological safety, and performance effects. *Production and Operations Management, 23*(8), 1352–1369.

Beth, S., Burt, D. N., Copacino, W., Gopal, C., Lee, H. L., Lynch, R. P., & Morris, S. (2003). Supply chain challenges – Building relationships. *Harvard Business Review, 81*(7), 64–73.

Birnbaum, P. H., Rossini, F. A., & Baldwin, D. R. (Eds.). (1990). *International research management: Studies in interdisciplinary methods from business, government, and academia*. New York: Oxford University Press.

Capra, F. (2002). *The hidden connections*. London: HarperCollins.

Carrubbo, L. (Ed.). (2013). *La Co-creazione di valore nelle destinazioni turistiche*. Rome: RIREA.

Carrubbo, L., Iandolo, F., & Pitardi, V. (2015). The viable decision maker for CAS survival: Between change and adaptation. In: E. Gummesson, C. Mele, & F. Polese (Eds.), *System theory and service science: Integrating three perspectives in a new service agenda*. Naples, 9–12 June 2015.

Cavaleri, S., Firestone, J., & Reed, F. (2012). Managing project problem-solving patterns. *International Journal of Managing Projects in Business, 5*(1), 125–145.

Checkland, P. (1997). *Systems thinking, systems practice*. Chichester: Wiley.

Checkland, P. (1999). Systems thinking. In W. Currie & B. Galliers (Eds.), *Rethinking management information systems: An interdisciplinary perspective*. Oxford: Oxford University Press.

Chen, Z., Dahlgaard-Park, S. M., & Yu, L. (2014). Service quality management and ecosystem theory. *Total Quality Management and Business Excellence, 25*(9-10), 1190–1205.

Choi, C. J., & Hilton, B. (2005). Knowledge resources: Critical systems thinking, viable system model and 'gifts'. *Systems Research and Behavioral Science, 22*(6), 561–564.

Christopher, W. F. (2007). *Holistic management: Managing what matters for company success*. Hoboken: Wiley Interscience.

Chronéer, D., & Backlund, F. (2015). A holistic view on learning in project-based organizations. *Project Management Journal, 46*(3), 61–74.

Chroneer, D., & Mirijamdotter, A. (2009). Systems thinking benefits in supply change management: An illustration of the viable systems model in a supply chain. *International Journal of Intelligent Systems Technologies and Applications, 6*(3–4), 227–248.

Conroy, D. M., & Allen, W. (2010). Who do you think you are? An examination of how systems thinking can help social marketing support new identities and more sustainable living patterns. *Australasian Marketing Journal (AMJ), 18*(3), 195–197.

Conti, T. (2010). System thinking in quality management. *The TQM Journal, 22*(4), 352–368.

Costello, K., Crawford, L., Bentleym, L., & Pollack, J. (2002). Connecting soft systems thinking with project management practice: An organizational change case study. In Ragsdell, et al. (Eds.), *Systems theory and practice in the knowledge age* (pp. 47–54). New York: Kluwer Academic.

De Santo, M., Pietrosanto, A., Napoletano, P., & Carrubbo, L. (2011). Knowledge based service systems. In E. Gummesson, C. Mele, & F. Polese (Eds.), *System theory and service science: Integrating three perspectives in a new service agenda*. Giannini, Naples, 14–17 June 2011.

Demirkan, H., Kauffman, R. J., Vayghan, J. A., Fill, H. G., Karagiannis, D., & Maglio, P. P. (2008). Service-oriented technology and management: Perspectives on research and practice for the coming decade. Perspectives on the Technology of Service Operations. *Electronic Commerce Research and Applications, 7*(4), 356–376.

Demirkan, H., Spohrer, J., & Krishna, V. (2011a). *Service systems implementation*. New York: Springer.

Demirkan, H., Spohrer, J., & Krishna, V. (2011b). *The science of service systems*. New York: Springer.

Donald, N. (2010). Systems thinking, complexity theory and transnational management. *Otago Management Graduate Review, 8*, 21–33.

Emery, F. E., & Trist, E. L. (1960). Socio-technical systems. In F. E. Emery (Ed.), *Systems thinking*. Middlesex: Penguin.

Espejo, R. (1999). Seeing systems: Overcoming organisational fragmentation. In A. M. Castell, A. J. Gregory, G. A. Hindle, M. E. James, & G. Ragsdell (Eds.), *Synergy matters: Working with systems in the 21st century*. New York: Kluwer Academic.

Espejo, R., & Harnden, R. J. (1989). *The viable system model*. London: John Wiley.

Espinosa, A., Harnden, R., & Walker, J. A. (2008). A complexity approach to sustainability – Stafford Beer revisited. *European Journal of Operational Research, 187*(2), 636–651.
Ganzert, C., Martinelli, D. P., & Delai, I. (2012). Intelligence systems methodology: A systemic approach to the organizational intelligence function. *Knowledge Management Research and Practice, 10*(2), 141–152.
Golinelli, G. M. (2000). *L'approccio sistemico al governo dell'impresa. L'impresa sistema vitale* (I ed.). Padova: CEDAM.
Golinelli, G. M. (2005). *L'approccio sistemico al governo dell'impresa. L'impresa sistema vitale*. Padova: CEDAM.
Golinelli, G. M. (2010). *Viable systems approach (VSA). Governing business dynamics*. Padova: Kluwer Cedam.
Guerreiro, E. P., Neto, P. L. D. O. C., & Moreira Filho, U. M. (2014). Multidisciplinary management: Model of excellence in the management applied to products and services. In *IFIP International Conference on Advances in Production Management Systems* (pp. 27–34), September 2014. Heidelberg: Springer.
Haines, S. (2000). *The systems thinking approach to strategic planning and management*. New York: St. Lucie Press.
Hitchins, D. K. (2003). *Advanced systems thinking, engineering and management*. Norwood, MA: Artech House.
Holweg, M., & Pil, F. K. (2008). Theoretical perspectives on the coordination of supply chains. *Journal of Operations Management, 26*(3), 389–406.
Jackson, M. (2003). *Systems thinking: Creative holism for managers*. Chichester: Wiley.
Jackson, M. (2009). Fifty years of systems thinking for managers. *Journal of the Operational Research Society, 60*(1), 24–32.
Kamppinen, M., Vihervaara, P., & Aarras, N. (2008). Corporate responsibility and systems thinking. Tools for balanced risk management. *International Journal of Sustainable Society, 1*, 158–171.
Karppinen, H., Seppänen, K., & Huiskonen, J. (2014). Identifying product and process configuration requirements in a decentralised service delivery system. *International Journal of Services and Operations Management, 17*(3), 294–310.
Katz, D., & Kahn, R. L. (1966). *The social psychology of organization*. New York: Wiley.
Katz, D., & Kahn, R. L. (1978). *The social psychology of organizations* (2nd ed.). New York: Wiley.
Layton, R. A. (2007). Marketing systems: A core macromarketing concept. *Journal of Macromarketing, 27*(3), 227–242.
Layton, R. A. (2011). Towards a theory of marketing systems. *European Journal of Marketing, 45*(1-2), 259–276.
Lee, L. S., & Green, E. (2015). Systems thinking and its implications in enterprise risk management. *Journal of Information Systems, 29*(2), 195–210.
Luhmann, N. (1990). *Soziale Sisteme. Grundriß einer allgemeinen Theorie*. Frankfurt: Suhrkamp Verla g.
Lusch, R., Vargo, S., & Tanniru, M. (2009). Service, value networks and learning. *Journal of the Academy of Marketing Science, 38*(1), 19–31.
Lyneis, J. M., & Ford, D. N. (2007). System dynamics applied to project management: A survey, assessment, and directions for future research. *System Dynamics Review, 23*(2-3), 157–189.
Maani, K. E., & Maharaj, V. (2004). Links between systems thinking and complex decision making. *System Dynamics Review, 20*(1), 21–48.
Maturana, H. R., & Varela, F. J. (1975). Autopoietic systems. BLC Report 9. University of Illinois.
Maull, M., Geraldi, J., & Johnston, R. (2012). Service supply chain: A customer perspective. *Journal of Supply Chain Management, 48*(4), 72–86.
Mawby, D., & Stupples, D. (2002). *Systems thinking for managing projects*. IEMC-2002: IEEE International Engineering Management Conference (Vols. I and II, pp. 344–349). Proceedings. Piscataway, NJ: IEEE.

Mele, C., Pels, J., & Polese, F. (2010). A brief review of systems theories and their managerial applications. *Service Science, 2*(1/2), 126–135.

Mingers, J., & White, L. (2010). A review of the recent contribution of systems thinking to operational research and management science. *European Journal of Operational Research, 207*, 1147–1161.

Moon, S. A., & Kim, D. J. (2005). Systems thinking ability for supply chain management. *Supply Chain Management – An International Journal, 10*(5), 394–401.

Murad, R. S. A., & Cavana, R. Y. (2012). Applying the viable system model to ICT project management. *International Journal of Applied Systemic Studies, 4*(3), 186–205.

Mutingi, M., & Mbohwa, C. (2014). Understanding sustainability in healthcare systems: A systems thinking perspective. In *2014 I.E. International Conference on Industrial Engineering and Engineering Management* (pp. 597–601), December 2014.

Napoletano, P., & Carrubbo, L. (2011). Becoming smarter: Towards a new generation of services systems. In: *Proceedings of 1st Cooper Link Workshop*, Vol. 3(3), August 2011.

Ndlela, M. N. (2014). Critical success factors for effective knowledge sharing: Integrating intra-organizational communication and KM tools. *European Conference on Knowledge Management, 2*, 724.

O'Donnell, E. (2005). Enterprise risk management: A systems-thinking framework for the event identification phase. *International Journal of Accounting Information Systems, 6*(3), 177–195.

Pagani, M., & Otto, P. (2013). Integrating strategic thinking and simulation in marketing strategy: Seeing the whole system. *Journal of Business Research, 66*(9), 1568–1575.

Paina, L., Bennett, S., Ssengooba, F., & Peters, D. H. (2014). Advancing the application of systems thinking in health: Exploring dual practice and its management in Kampala, Uganda. *Health Research Policy and Systems, 12*(1), 1.

Parsons, T. (1965). *Il Sistema Sociale*. Torino Edizioni: Comunità.

Paucar-Caceres, A., & Pagano, R. (2009). Systems thinking and the use of systemic methodologies in knowledge management. *Systems Research and Behavioral Science, 26*(3), 343–355.

Pentland, D., Forsyth, K., Maciver, D., Walsh, M., Murray, R., & Irvine, L. (2014). Enabling integrated knowledge acquisition and management in health care teams. *Knowledge Management Research and Practice, 12*(4), 362–374.

Polese, F., & Carrubbo, L. (2016). *Eco-sistemi di servizio in Sanità*. Torino: Giappichelli Editore.

Polese, F., Carrubbo, L., & Russo, G. (2010). Managing business relationships between service culture and viable systems approach. *Esperienze di Impresa, 2*, 121–144.

Polese, F., Carrubbo, L., & Di Nauta, P. (2012). Smart energy management – A viable systems approach perspective. *Business Systems Review, 1*(1), 202–224.

Powell, J. (2004). An introduction to systems theory: From hard to soft systems thinking in the management of complex organizations. In D. Kernick (Ed.), *Complexity and healthcare organization: A view from the street*. Boca Raton: CRC Press.

Puche, J., Ponte, B., Costas, J., Pino, R., & De La Fuente, D. (2016). Systemic approach to supply chain management through the viable system model and the theory of constraints. *Production Planning and Control, 27*(5), 421–430.

Rabaey, M. (2013). Complex adaptive systems thinking approach for intelligence base in support of intellectual capital management. In P. Ordòñez de Pablos, T. Tennyson, & J. Zhao (Eds.), *Intellectual capital strategy management for knowledge-based organizations* (pp. 122–141). Hershey, PA: Business Science Reference.

Rana, S., Crowe, M., & Usoro, A. (2013). Factors for knowledge sharing behaviour to develop trust. Professional Organisations Environment.

Reiman, T., Rollenhagen, C., Pietikäinen, E., & Heikkilä, J. (2015). Principles of adaptive management in complex safety-critical organizations. *Safety Science, 70*, 80–92.

Senge, P. M., & Sterman, J. D. (1992). Systems thinking and organizational learning: Acting locally and thinking globally in the organization of the future. *European Journal of Operational Research, 59*(1), 137–150.

Sheffield, J., Sankaran, S., & Haslett, T. (2012). Systems thinking: Taming complexity in project management. *On the Horizon, 20*(2), 126–136.
Shen, H., Wang, L., Xu, Q., Li, Y., & Liu, X. (2009). Toward a framework of innovation management in logistics firms: a systems perspective. *Systems Research and Behavioral Science, 26*(2), 297–309.
Sole, F., & Schiuma, G. (2010). Understanding organisational knowledge – Based value creation dynamics: A systems thinking approach. In P. Taticchi (Ed.), *Business performance measurement and management new contents, themes and challenges* (pp. 327–341). Berlin: Springer-Verlag.
Spohrer, J., Vargo, S. L., Maglio, P. P., & Caswell, N. (2008). The service system is the basic abstraction of service science. In *HICSS Conference* (pp. 104–104).
Sterman, J. D. (1994). Learning in and about complex systems. *System Dynamics Review, 10*(2/3), 291–330.
Tien, J. M., & Berg, D. (2003). A case for service systems engineering. *Journal of Systems Science and Systems Engineering, 12*(1), 13–38.
Turnbull, S. (2002). The science of corporate governance. *Corporate Governance-an International Review, 10*(4), 261–277.
Turnbull, J. E. (2002). *Process management and systems thinking for patient safety. The Business of Health Care: A Journal of Innovative Management Collection.* Salem, NH: Goal/QPC.
Urze, P., & Abreu, A. (2014). System thinking to understand networked innovation. In *Working Conference on Virtual Enterprises* (pp. 327–335), October 2014. Heidelberg: Springer.
Vargo, S. L., & Lusch, R. F. (2011). It's all B2B...and beyond: Toward a systems perspective of the market. *Industrial Marketing Management, 40*, 181–187.
Vargo, S. L., & Lusch, R. F. (2016). Institutions and axioms: an extension and update of service-dominant logic. *Journal of the Academy of Marketing Science, 44*(1), 5–23.
Vargo, S. L., Maglio, P. P., & Akaka, M. A. (2008). On value and value co-creation – A service systems and service logic perspective. *European Management Journal, 26*(3), 145–152.
Von Bertalanffy, L. (1956). General system theory. In F. E. Emery (Ed.), *General system.* Yearbook of the Society for the Advancement of General System Theory.
Von Bertalanffy, L. (1962). *Modern theories of development.* New York: Harper.
Von Foerster, H. (1981). *Observing systems.* Seaside: Inter Systems Publication.
Waliullah, M., & Schell, W. J. (2013). Toward improving healthcare: A meta-analysis of leadership approaches. In *International Annual Conference of the American Society for Engineering Management 2013*, ASEM in St. Paul, MN.
Weinberg, G. M. (2001). *An introduction to general systems thinking.* Dorset House Publishing Company, Incorporated; 25 Anv edition, April.
Wendorff, P. (2002). An essential distinction of agile software development processes based on systems thinking in software engineering management. In *Third International Conference on eXtreme Programming and Agile Processes in Software Engineering*, Alghero, Sardinia.
Wieland, H., Polese, F., & Vargo, S. L. (2012). Toward a service (eco)systems perspective on value creation. *International Journal of Service Science, Management, Engineering and Technology, 3*(3), 12–25.
Williams, T. (2016). Identifying success factors in construction projects: A case study. Source of the document. *Project Management Journal, 47*(1), 97–112.
Winter, M., & Checkland, P. (2003). Soft systems: A fresh perspective for project management. *Proceedings of the Institution of Civil Engineers-Civil Engineering, 156*(4), 187–192.
Wolf, P., Meissner, J. O., Nolan, T., Lemon, M., John, R., Baralou, E., & Seemann, S. (2011). Methods for qualitative management research in the context of social systems thinking. *Historical Social Research, 36*(1), 7–18.
Woodside, A. G. (2006). Advancing systems thinking and building microworlds in business and industrial marketing. *Journal of Business and Industrial Marketing, 21*(1), 24–29.
Xiang-yu, K., & Xiang-yang, L. (2007). A systems thinking model for innovation management: The knowledge management perspective. In *The 14th International Conference on Management Science & Engineering* (pp. 1499–1504). Harbin: IEEE Press.

MIX
Papier aus verantwortungsvollen Quellen
Paper from responsible sources
FSC® C105338

If you have any concerns about our products,
you can contact us on
ProductSafety@springernature.com

In case Publisher is established outside the EU,
the EU authorized representative is:
Springer Nature Customer Service Center GmbH
Europaplatz 3, 69115 Heidelberg, Germany

Printed by Libri Plureos GmbH
in Hamburg, Germany